U0349140

菠萝种质资源图谱（下册）

Pineapple Germplasm Resources Map (Volume 2)

◎ 张秀梅　陆新华　姚艳丽　刘胜辉　等　著

中国农业科学技术出版社

图书在版编目（CIP）数据

菠萝种质资源图谱. 下册 / 张秀梅等著. -- 北京：中国农业科学技术出版社，2024. 9. -- ISBN 978-7-5116-7047-2

Ⅰ. S668.3-64

中国国家版本馆 CIP 数据核字第 2024JQ2543 号

责任编辑 史咏竹 白 净
责任校对 马广洋
责任印制 姜义伟 王思文

出 版 者 中国农业科学技术出版社
北京市中关村南大街 12 号 邮编：100081
电 话 （010）82105169（编辑室）（010）82106624（发行部）
（010）82109709（读者服务部）
网 址 https://castp.caas.cn
经 销 者 各地新华书店
印 刷 者 北京地大彩印有限公司
开 本 170 mm × 240 mm 1/16
印 张 12
字 数 206 千字
版 次 2024 年 9 月第 1 版 2024 年 9 月第 1 次印刷
定 价 98.00 元

《菠萝种质资源图谱（下册）》
著者委员会

著者名单

主　　著　张秀梅　陆新华　姚艳丽　刘胜辉

副 主 著　黄炳钰　吴青松　贺军军　李川玲

参著人员　（以姓氏笔画为序）

井敏敏　付　琼　朱祝英　孙光明

孙伟生　杜丽清　杨玉梅　汪佳维

林文秋　高玉尧　夏　瑞

著者单位

中国热带农业科学院南亚热带作物研究所

农业农村部湛江菠萝种质资源圃

国家热带植物种质资源库

热带作物生物育种全国重点实验室

农业农村部热带果树生物学重点实验室

海南省菠萝种质创新与利用工程技术研究中心

国家重要热带作物工程技术研究中心—菠萝研发部

华南农业大学园艺学院

中国热带农业科学院三亚研究院

Pineapple Germplasm Resources Map（Volume 2）

Authors Board

The list of Authors

Author in Chief Zhang Xiumei, Lu Xinhua, Yao Yanli, Liu Shenghui

Associate Authors in Chief Huang Bingyu, Wu Qingsong, He Junjun, Li Chuanling

Authors (Sort by the strokes of the surnames in Chinese)

Jing Minmin, Fu Qiong, Zhu Zhuying, Sun Guangming, Sun Weisheng, Du Liqing, Yang Yumei, Wang Jiawei, Lin Wenqiu, Gao Yuyao, Xia Rui

Affiliation of the Authors

South Subtropical Crops Research Institute, Chinese Academy of Tropical Agricultural Sciences

Germplasm Repository of Pineapple (*Ananas comosus*) Zhanjiang City, Ministry of Agriculture and Rural Affairs

National Tropical Plants Germplasm Resource Center

National Key Laboratory for Tropical Crop Breeding

Key Laboratory of Tropical Fruit Biology, Ministry of Agriculture and Rural Affairs

Hainan Provincial Egineering Rsearch Center for Pineapple Germplasm Innovation and Utilization

Pineapple R&D Department, National Center of Important Tropical Crops Engineering and Technology Research

College of Horticulture, South China Agricultural University

Sanya Research Institute of Chinese Academy of Tropical Agricultural Sciences

本书资助项目：农业农村部农垦局项目“菠萝、香蕉、澳洲坚果、荔枝等南亚热带作物种质资源保护与利用”（项目编号：A120202）；国家重点研发计划课题“菠萝种质资源精准评价与基因发掘”（项目编号：2019YFD1000505）；广东省农业农村厅项目“岭南特色水果新品种区域示范及推广”（项目编号：44000022000000035430）；海南省重点研发专项“揭榜挂帅”项目“海南凤梨优质新品种选育与推广研究”。

This book was supported by projects of Agricultural Reclamation Bureau, Ministry of Agriculture and Rural Affairs (A120202), National Key Research and Development Program of China (2019YFD1000505), Agricultural and Rural Department of Guangdong Province (44000022000000035430), Hainan Province Science and Technology Special Fund.

本书采集图片资料的田间工作操作地点：农业农村部湛江菠萝种质资源圃。

The field operation location for collecting image materials in this book is the Germplasm Repository of Pineapple (*Ananas comosus*) Zhanjiang City, Ministry of Agriculture and Rural Affairs.

本书采集图片资料及统计性状过程中各品种菠萝定植时间：2021—2022 年。

During the process of collecting image materials and statistical traits in this book, various varieties of pineapple were planted from 2021 to 2022.

目 录

Contents

第一章
菠萝种质资源遗传多样性图谱

Chapter 1
Genetic Diversity Map of Pineapple Germplasm Resources

菠萝［*Ananas comosus* (L.) Merr.］是凤梨科（Bromeliaceae）凤梨属（*Ananas* Merr.）多年生草本植物。菠萝原产于南美洲，1493 年被欧洲人首次发现，后来推广至世界各地，现广泛分布于热带和亚热带地区。菠萝果实风味独特、香气怡人、营养丰富，可鲜食，亦可制成罐头、果汁或用于提取菠萝蛋白酶等，是世界四大热带水果（香蕉、菠萝、荔枝和杧果）之一，具有重要的经济价值。我国最早引进菠萝是在 16 世纪末，迄今已有 400 多年历史。菠萝现已发展成为我国热区重要的经济果品之一，主要分布在广东、海南、广西、云南、台湾等省区。中国热带农业科学院南亚热带作物研究所从 2003 年起引入菠萝，目前收集保存 150 多份种质资源，种植于农业农村部湛江菠萝种质资源圃内。本图谱仅对圃内保存的有代表性的菠萝种质资源的植物学性状、农艺性状和品质性状等进行了描述。

不同品种菠萝种质资源在植株姿态、冠芽特征、冠芽外形、冠芽叶刺、叶片彩带、叶片叶刺、叶刺生长方向等方面具有一定的差异。植株姿态有直立（夹角≥80°）、开张（40°≤夹角<80°）、匍匐（夹角<40°）；冠芽特征有单冠芽、双冠芽、多冠芽、单小冠芽（冠芽高度小于果体高度的 1/2）；冠芽外形有椭圆形、圆柱形、圆锥形、扇形、喇叭形、其他形态；冠芽叶刺状态分为光滑无刺、部分叶缘有刺、叶尖有刺、全缘有刺；叶片着生姿态分为直立、开张、平展、下垂；叶片颜色有淡绿 / 绿色、绿色带黄色斑纹、绿色带紫红色斑纹、暗红色、浅紫 / 紫红色、深紫 / 暗紫红色、其他颜色等；叶片彩带状态分为无、两侧、中央；叶片叶刺分布状态有光滑无刺、仅少量分布在叶尖处、仅少量且无规律地分布在叶缘处、较多且无规律地分布在叶缘处、布满整个叶缘；叶刺生长方向有向上顺生、向上顺生与向下倒生兼备；叶刺密度有少量 / 稀疏（密度≤1 枚 /cm）、中度（1 枚 /cm<密度<3 枚 /cm）、很多 / 密集（密度≥3 枚 /cm）等类型。

Pineapple [*Ananas comosus* (L.) Merr.] belonging to the genus *Ananas* of the family Bromeliaceae, is a perennial herbaceous plant. Pineapple was originated from South America. It was first discovered by Europeans in 1493. Later, pineapple was spread to many other parts of the world and now it is widely distributed in the tropical and subtropical areas. Pineapple fruit has rich nutrition with unique flavors and a pleasant aroma, and is usually consumed as fresh fruit or canned products, it can also be used to produce juices or bromelains. Pineapple is one of the four most important tropical fruits (banana, pineapple, litchi and mango) in the world and it has an essential economic value. The earliest introduction of pineapple to China was at the end of the 16th Century and now it has more than 400 years of history here. Currently, pineapple has become one of principal economic fruit in tropical regions of China, being mainly distributed in Guangdong, Hainan, Guangxi, Yunan, Taiwan, and other places. Since 2003, South Subtropical Crops Research Institute (SSCRI) of Chinese Academy of Tropical Agricultural Sciences (CATAS) has introduced its first pineapple germplasm, and now more than 150 pineapple accessions are collected and conserved in Germplasm Repository of Pineapple (*Ananas comosus*) Zhanjiang City, Ministry of Agriculture and Rural Affairs. This book introduces some typical pineapple germplasm resources conserved in the germplasm repository and their botanical, agronomic and quality characteristics, etc.

Pineapple germplasm resources possess rich diversity in the aspects of their plant postures, crown bud characters, crown bud shape, crown bud spines, leaf ribbons, leaf spines, leaf spine growth directions, etc. The plant posture of pineapple can be either upright (angle≥80°), spreading (40°≤angle<80°) or procumbent (angle<40°). They could have either a single crown bud, a double crown bud, a multiple crown bud or a single small crown bud (the height of the crown bud is less than half of the height of the fruit body). The crown bud can be either ellipsoid, cylindrical, conical, fan-shaped, trumpet shaped, etc. The crown bud leaf has smooth margin without spines, partly margin with spines, tip with spines or overall margins with spines. The leaf posture can be either erect, spreading, expanding or droopy. Leaf color can be green/light green, green with yellow stripe, green with purplish red stripe, dark red, light purple/purplish red, dark purple/dark purplish red, etc. Some varieties do not have leaf ribbon but others have it in the middle or on both sides of the leaf. The characteristics of leaf spines vary from one variety to another: smooth margins without spines, few spines distributing at the leaf tips, few spines irregularly distributing at the leaf margins, many spines regularly distributing at the leaf margins, spines regularly distributing at both sides of the leaf margins, or

在菠萝花序上的小花开放盛期，花瓣开张状态有微开、半张开、张开；按照花瓣间是否重叠及其状态，花冠形态有旋转状、覆瓦状；苞片边缘状态有锯齿状、光滑。

菠萝果实具有丰富的遗传多样性，按果实形状可分为圆台形 / 方形、（近）圆球形、圆筒形、长圆筒形、圆锥形、长圆锥形、其他；按未成熟果实果皮颜色可分为银绿色、淡绿 / 绿色、暗绿色、暗墨绿色、淡黄 / 黄绿色、浅红 / 粉红 / 橙红 / 浅褐色、红色、红色中略显紫色、暗紫红色、紫色、蓝紫色等；按成熟果实果皮颜色可分为绿色、黄色，带绿斑、暗黄 / 深黄色、亮黄 / 淡黄色、金黄 / 鲜黄色等；按果基形状可分为平、弧形、突起；按果顶形状可分为平、浑圆、钝圆、尖圆；果颈可分为无颈、有颈；按果眼外观可分为扁平或微凹、微隆起、突起 / 隆起；按果眼深度可分为深（深度≥1.2 cm）、较深（1.1 cm≤深度＜1.2 cm）、较浅（0.9 cm≤深度＜1.1 cm）和浅（深度＜0.9 cm）；按果眼排列方式可分为左旋、右旋、其他类型；按果实底部着生的果瘤可分为无、少（果瘤数 1～2 个）、多（果瘤数≥3 个）；按果实小果可分为不可剥离、可剥离；按果肉颜色可分为白色、奶油色、淡黄色、黄色、金黄色、深黄色、橙黄色等；按果实香味可分为无香、清香 / 微香、芳香、极香 / 浓香等；按果肉风味可分为浓甜、清甜、甜酸、酸、酸甜、极酸、香甜、微香甜、微涩等；按果肉质地可分为滑、脆/爽脆、粗糙。按果实成熟特性分为特早熟（成熟期≤60 d）、早熟（60 d＜成熟期＜80 d）、中熟（80 d≤成熟期＜100 d）、晚熟（≥100 d）等；根据菠萝果实第一次采收至最后一次采收之间所持续的天数，成熟期的一致性可分为一致（相差天数≤5 d）、基本一致（5 d＜相差天数＜15 d）、不一致（相差天数≥15 d）；按产量特性可分为丰产（产量≥40 t/hm^2）、一般产量（30 t/hm^2≤产量＜40 t/hm^2）、低产（产量＜30 t/hm^2）。

spines covering the entire leaf margins. The leaf spines are either antrorse or both antrorse and retrorse at the same time. Spine density can be sparse (density≤1/cm), medium (1/cm<density<3/cm), or many/condensed (density≥3/cm), etc.

During the full-blooming stage, petals of florets in an inflorescence can be slightly open, half open, or fully open. According to whether the petals overlap and their state, the corolla morphology can be classified into either contorted or imbricate. The margins of bract can be either serrated or smooth.

The pineapple fruit presents a rich genetic diversity. The fruit shape can be either truncated cone/square, (near) spherical, cylindrical, long cylindrical, conical, long conical, pyriform, etc. The immature fruit peel can be silver green, light green/green, dark green, dark blackish green, light yellow/yellowish green, light red/candy pink/orange red/light brown, red, red with slight purple, violet red, purple, blue purple, etc. In addition, the mature fruit peel can be green, yellow with green stripe, dark yellow/deep yellow, bright yellow/light yellow, golden yellow/vivid yellow, etc. Fruit base can be flat, curved or protrusive. Fruit top can be flat, perfectly round, bluntly round or sharply round. Some pineapples show obvious fruit neck, others have null. Fruit eyes (fruitlets) can be flat, slightly concave, slightly embossed, or convex/embossed. The fruit eyes depth can be deep (depth≥1.2 cm), relatively deep (1.1 cm≤depth<1.2 cm), relatively shallow (0.9 cm≤depth<1.1 cm) or shallow (depth<0.9 cm). The fruit eyes arrangement can be levorotation, dextrorotation or other type. Some varieties do not have fruit tumors on the bottom of the fruit, but others have few (1–2) or many (≥3) fruit tumors. Some fruitlets can be easily separated but others can not. Flesh color can be white, cream, pale yellow, yellow, golden yellow, deep yellow, orange, etc. Some fruits give off no scent but many other fruits send out faint/slight scent, sweet fragrance, or even strong/extreme aroma. Fruit flavour can be strong sweet, mildly sweet, sweet-sour, sour, sour-sweet, extreme sour, fragrant and sweet, slightly fragrant and sweet, astringent, etc. The flesh texture can be smooth, crisp/crunchy or coarse. Fruit ripening characteristics can be divided into extremely early mature (mature period≤60 d), early mature (60 d<mature period<80 d), medium mature (80 d≤mature period<100 d) or late mature (mature period≥100 d). According to the duration days between the first and the last harvest of fruit, the maturity consistency can be uniform (duration of days≤5 d), basically uniform (5 d<duration of days<15 d) or nonuniform (duration of days≥15 d). The fruit yield characteristics can be productive (production≥40 t/hm^2), normal level (30 t/hm^2≤production<40 t/hm^2) or low (production<30 t/hm^2).

植株姿态 Plant posture

直立
Upright

开张
Spreading

匍匐
Procumbent

冠芽特征　Crown bud characteristic

单冠芽
Single

双冠芽
Double

多冠芽
Multiple

单小冠芽
Single small

冠芽外形 Crown bud shape

椭圆形
Ellipsoid

圆柱形
Cylindrical

圆锥形
Conical

扇形
Fan-shaped

喇叭形
Trumpet shaped

冠芽叶刺　Leaf spines of crown bud

光滑无刺
Absent

部分叶缘有刺
Part leaf margin with spines

叶尖有刺
Leaf tip with spines

全缘有刺
Overall leaf margin with spines

叶片着生姿态　Leaf posture

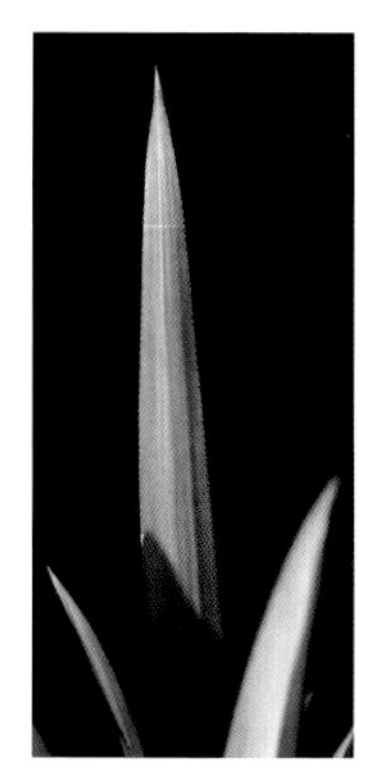

竖直
Erect

开张
Spreading

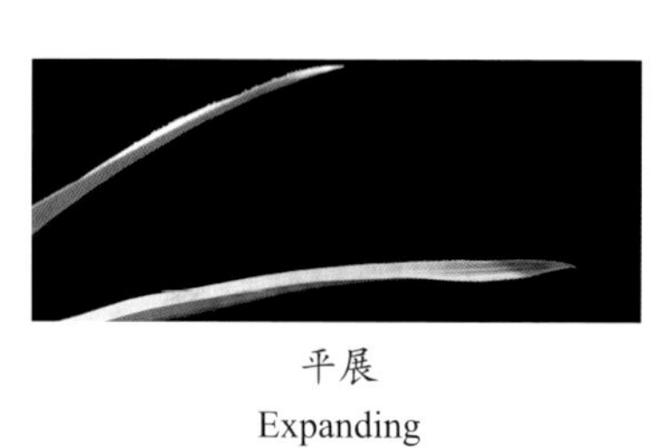

平展
Expanding

下垂
Droopy

叶片彩带状态 Leaf ribbon

无
Absent

两侧
Both sides

中央
Middle

叶片叶刺分布状态 Distribution of leaf spines

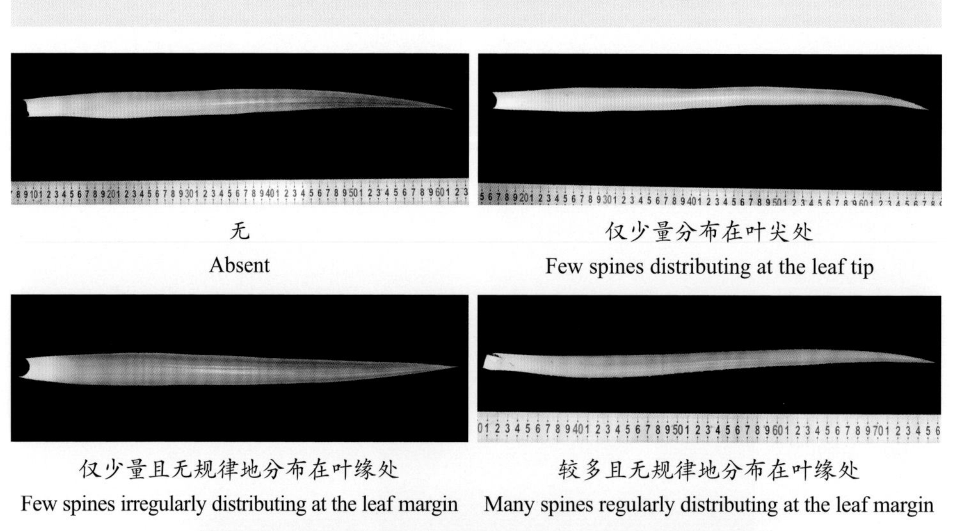

无
Absent

仅少量分布在叶尖处
Few spines distributing at the leaf tip

仅少量且无规律地分布在叶缘处
Few spines irregularly distributing at the leaf margin

较多且无规律地分布在叶缘处
Many spines regularly distributing at the leaf margin

布满整个叶缘
Spines covering the entire leaf margin

叶刺生长方向　Growth direction of leaf spines

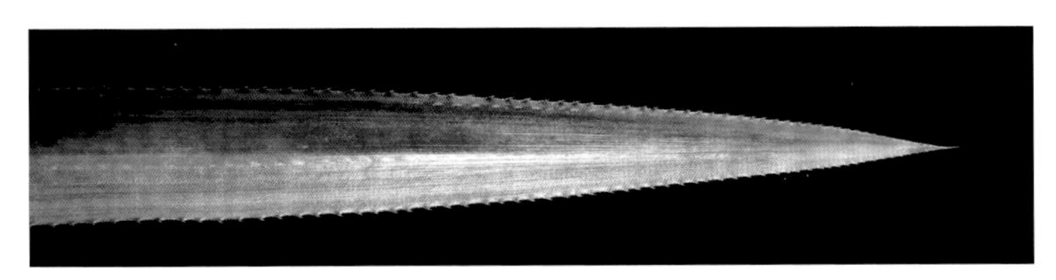

向上顺生
Antrorse

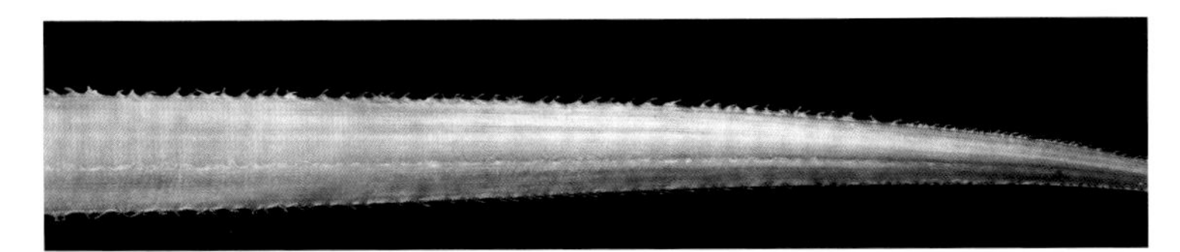

向上顺生与向下倒生兼备
Both antrorse and retrorse

叶刺密度　Density of leaf spines

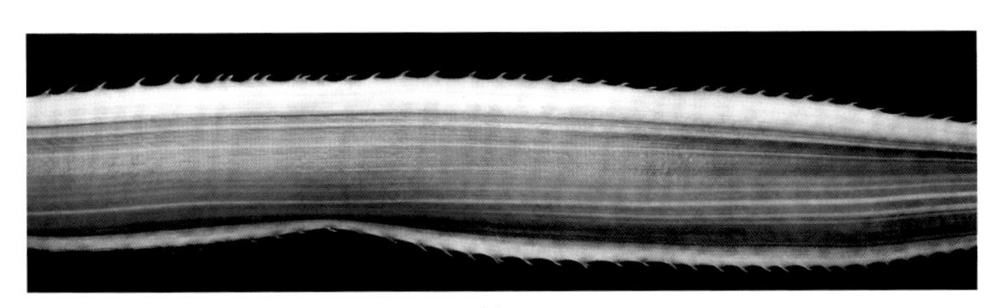

稀疏
Sparse

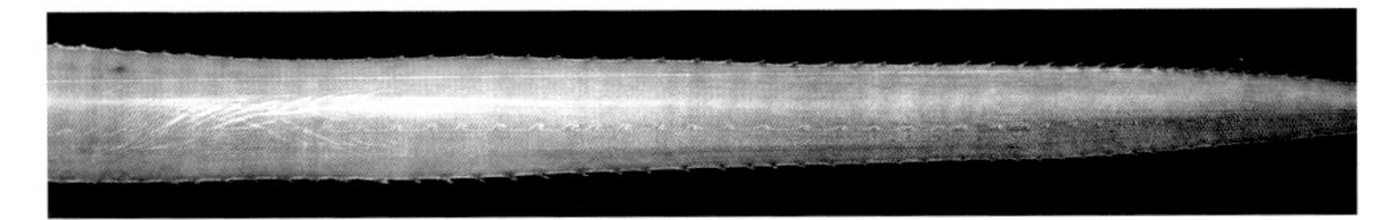

中度
Medium

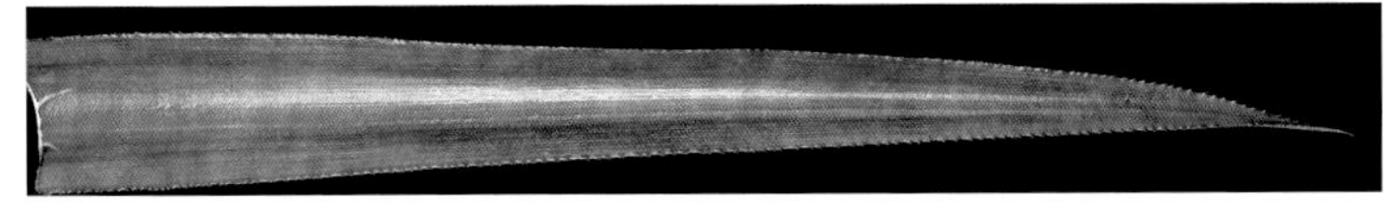

密集
Condensed

花瓣开张状态 State of petal opening

微开
Slightly open

半张开
Half open

张开
Fully open

花冠形态 Corolla morphology

旋转状
Contorted

覆瓦状
Imbricate

苞片边缘状态　Characteristics of bract margin

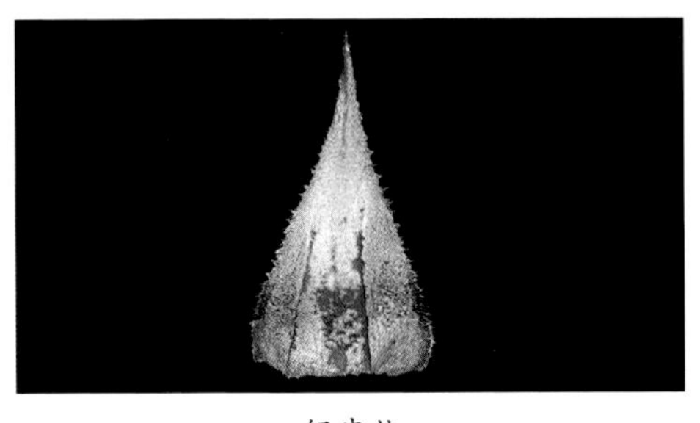

锯齿状
Serrated

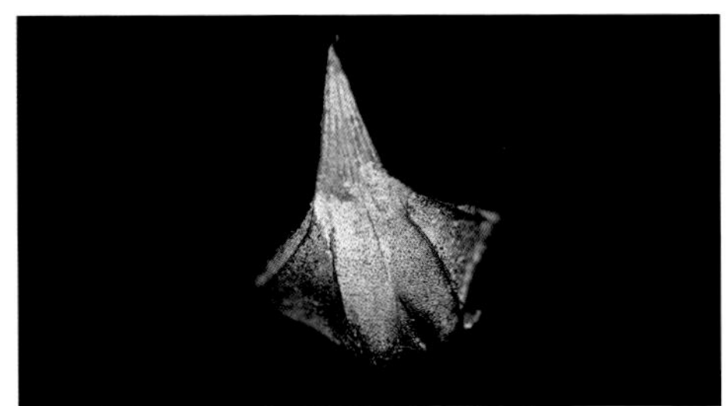

光滑
Smooth

果实形状　Fruit shape

圆台形
Truncated cone

（近）圆球形
(near) Spherical

圆筒形
Cylindrical

长圆筒形
Long cylindrical

圆锥形
Conical

长圆锥形
Long conical

果基形状 Fruit base shape

平 Flat　　弧形 Curved　　突起 Protrusive

果顶形状 Fruit top shape

平 Flat　　浑圆 Perfectly round　　钝圆 Bluntly round　　尖圆 Sharply round

果颈 Fruit neck

无
Absent

有
Present

果眼外观 Fruit eyes exterior

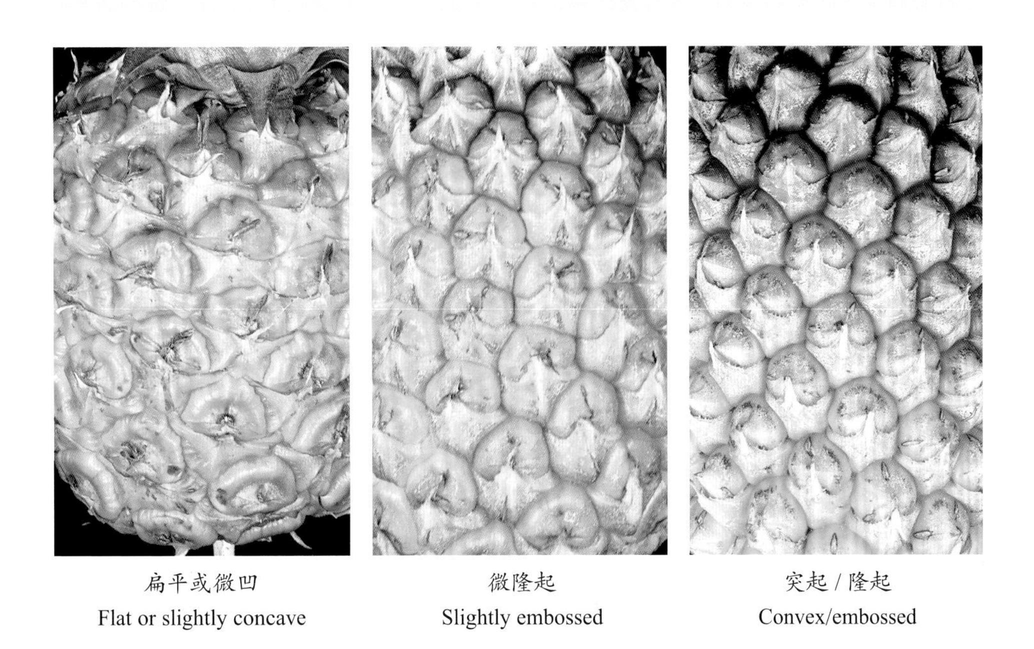

扁平或微凹
Flat or slightly concave

微隆起
Slightly embossed

突起 / 隆起
Convex/embossed

果眼排列方式 Fruit eyes arrangement

左旋
Levorotation

右旋
Dextrorotation

其他类型
Other type

果实底部着生的果瘤 Fruit tumors on the bottom of the fruit

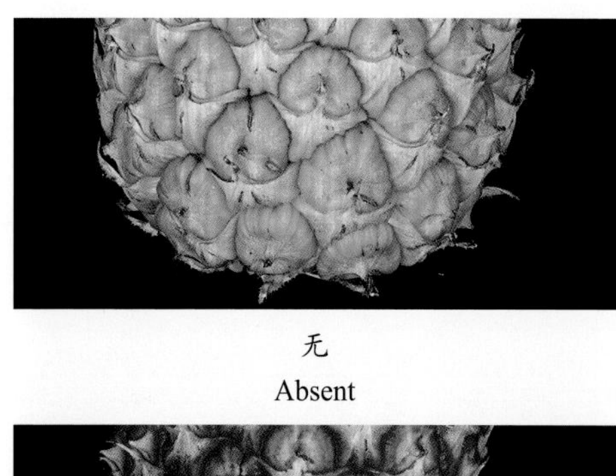

无
Absent

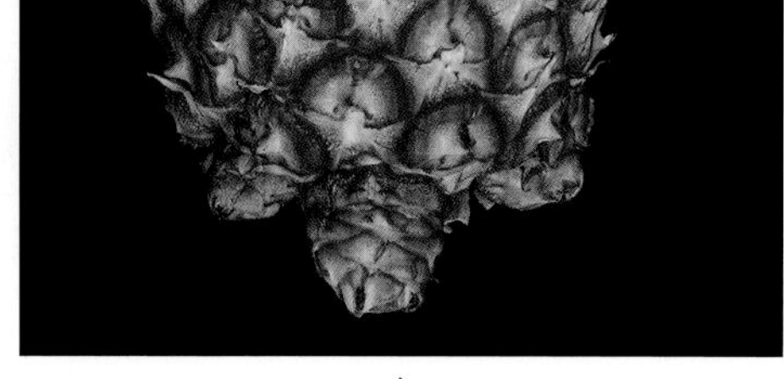

多
Many

少
Few

果肉颜色　Flesh color

白色
White

奶油色
Cream

淡黄色
Pale yellow

黄色
Yellow

金黄色
Golden yellow

深黄色
Deep yellow

橙黄色
Orange

第二章
国内菠萝种质资源

Chapter 2
Pineapple Germplasm Resources in China

巴厘

编号：ACCESSION000004
种质名称：巴厘
原产地：不详
资源类型：国内收集
主要用途：鲜食或加工
种质来源地：2007 年中国热带农业科学院南亚热带作物研究所从广东省收集
植株姿态：开张
定植期：2022 年 1 月中旬（湛江，余同）
现红期：2023 年 3 月上旬（自然果，余同）
营养生长期：约 425 d
初花期：2023 年 4 月上旬
花开放时间：约 22 d
成熟期：2023 年 7 月下旬至 8 月上旬
果实发育期：约 115 d
果实形状：长圆筒形
单果质量：915.9 g
纵径：14.0 cm
横径：11.0 cm
果形指数：1.3
未成熟果实果皮颜色：淡绿 / 绿色
成熟果实果皮颜色：亮黄 / 淡黄色
果颈：无
果基：平
果顶：平
果实小果能否剥离：可剥离
果眼外观：微隆起
果眼深度：较深
果眼排列方式：左旋
果瘤：多
果肉颜色：淡黄色
果实香味：清香 / 微香
果实风味：酸甜
果肉质地：脆 / 爽脆
果实外观综合评价：优
果实品质综合评价：好
果肉可溶性固形物含量：16.5%
果实成熟特性：中熟
成熟期的一致性：基本一致
丰产性：一般

Comte de Paris

Code: ACCESSION000004
Name: Comte de Paris
Place of origin: unknown
Resource type: domestically collected variety
Main ways of consumption: fresh or processing
Material resource: South Subtropical Crops Research Institute of Chinese Academy of Tropical Agricultural Science (SSCRI, CATAS, the same below) collected from Guangdong Province in 2007
Plant posture: spreading
Planting time: mid-January in 2022
Open heart stage: early March in 2023
Vegetative growth period: approx. 425 d
Initial flowering time: early April in 2023
Flowering period: approx. 22 d
Fruit maturation time: late July to early August in 2023
Fruit developing period: approx. 115 d
Fruit shape: long cylindrical
Single fruit weight: 915.9 g
Longitudinal diameter: 14.0 cm
Transverse diameter: 11.0 cm
Fruit shape index: 1.3
Immature fruit peel color: light green/green
Mature fruit peel color: bright yellow/light yellow
Fruit neck: absent
Fruit base: flat
Fruit top: flat
Fruitlets adhesion situation: separable
Fruit eyes appearance: slightly embossed
Fruit eyes depth: relatively deep
Fruit eyes arrangement: levorotation
Fruit tumors: many
Flesh color: pale yellow
Fruit aroma: faint/slight scent
Fruit flavour: sour-sweet
Flesh texture: crisp/crunchy
Fruit appearance comprehensive evaluation: excellent
Fruit quality comprehensive evaluation: good
Soluble solids content in flesh: 16.5%
Fruit ripening characteristics: medium mature
Maturity consistency: basically uniform
Productivity: normal level

植株
Plant

现红
Open heart

花
Flower

花序
Inflorescence

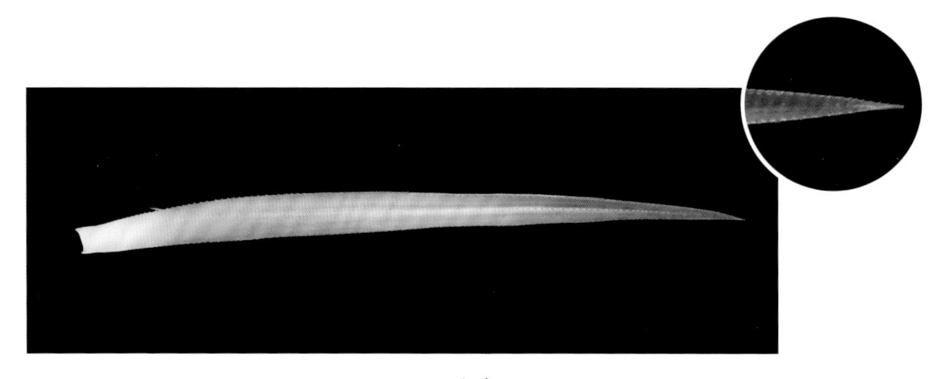

叶片
Leaf

带冠芽果
Fruit with crown bud

冠芽叶刺
Leaf spines in crown bud

果实
Fruit

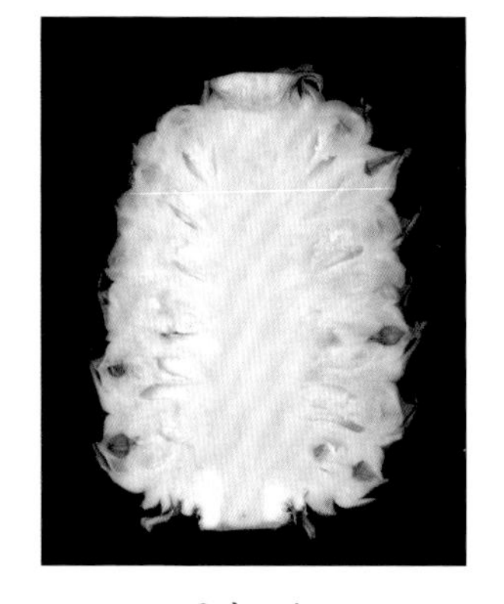

果实纵切
Fruit longitudinal section

果实横切
Fruit transverse section

珍珠

编号：ACCESSION000002
种质名称：珍珠
原产地：我国台湾
资源类型：国内收集
主要用途：鲜食或加工
种质来源地：2006年中国热带农业科学院南亚热带作物研究所从海南收集
植株姿态：开张
定植期：2021年2月中旬
现红期：2023年3月上中旬
营养生长期：约750 d
初花期：2023年4月上中旬
花开放时间：约25 d
成熟期：2023年8月上旬
果实发育期：约115 d
果实形状：长圆筒形
单果质量：895.8 g
纵径：12.2 cm
横径：10.6 cm
果形指数：1.2
未成熟果实果皮颜色：银绿色
成熟果实果皮颜色：暗黄 / 深黄色
果颈：无
果基：平
果顶：浑圆
果实小果能否剥离：可剥离
果眼外观：微隆起
果眼深度：浅
果眼排列方式：左旋
果瘤：无
果肉颜色：淡黄色
果实香味：芳香
果实风味：酸甜
果肉质地：滑
果实外观综合评价：优
果实品质综合评价：中
果肉可溶性固形物含量：11.1%
果实成熟特性：晚熟
成熟期的一致性：一致
丰产性：一般

Pearl

Code: ACCESSION000002
Name: Pearl
Place of origin: Chinese Taiwan
Resource type: domestically collected variety
Main ways of consumption: fresh or processing
Material resource: SSCRI, CATAS collected from Hainan Province in 2006
Plant posture: spreading
Planting time: mid-February in 2021
Open heart stage: early-to-mid March in 2023
Vegetative growth period: approx. 750 d
Initial flowering time: early-to-mid April in 2023
Flowering period: approx. 25 d
Fruit maturation time: early August in 2023
Fruit developing period: approx. 115 d
Fruit shape: long cylindrical
Single fruit weight: 895.8 g
Longitudinal diameter: 12.2 cm
Transverse diameter: 10.6 cm
Fruit shape index: 1.2
Immature fruit peel color: silver green
Mature fruit peel color: dark yellow/deep yellow
Fruit neck: absent
Fruit base: flat
Fruit top: perfectly round
Fruitlets adhesion situation: separable
Fruit eyes appearance: slightly embossed
Fruit eyes depth: shallow
Fruit eyes arrangement: levorotation
Fruit tumors: absent
Flesh color: pale yellow
Fruit aroma: sweet fragrance
Fruit flavour: sour-sweet
Flesh texture: smooth
Fruit appearance comprehensive evaluation: excellent
Fruit quality comprehensive evaluation: medium
Soluble solids content in flesh: 11.1%
Fruit ripening characteristics: late mature
Maturity consistency: uniform
Productivity: normal level

植株
Plant

现红
Open heart

花
Flower

花序
Inflorescence

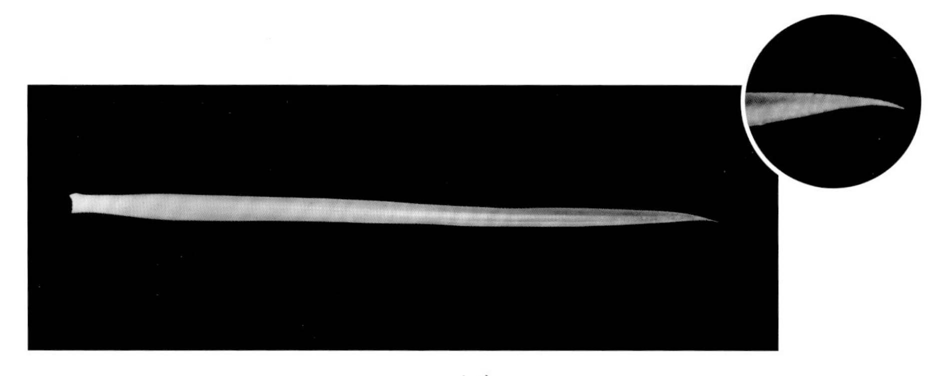

叶片
Leaf

带冠芽果
Fruit with crown bud

冠芽叶刺
Leaf spines in crown bud

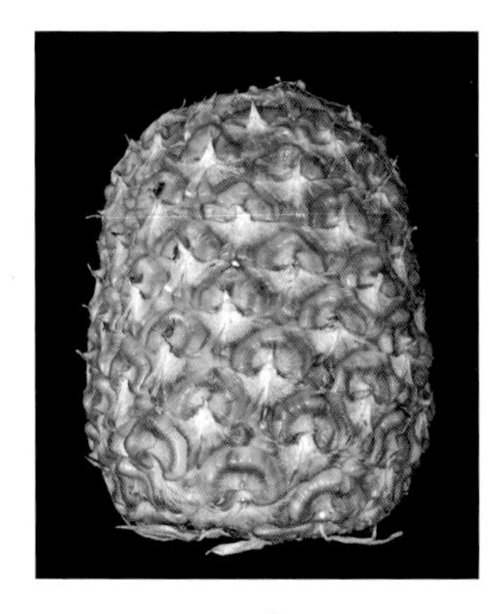

果实
Fruit

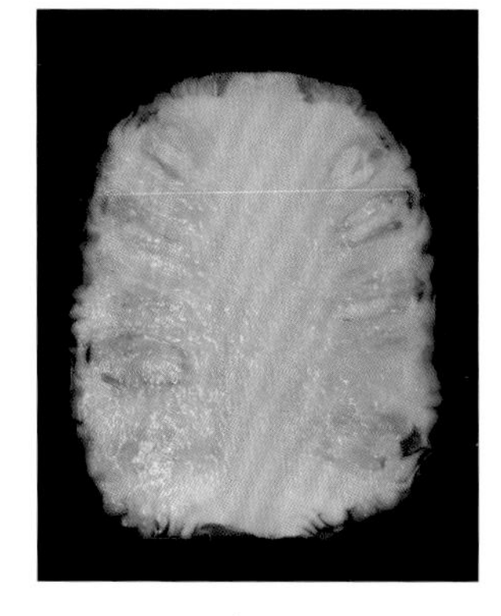

果实纵切
Fruit longitudinal section

果实横切
Fruit transverse section

台农 16 号

编号：ACCESSION000020
种质名称：台农 16 号，又名甜蜜蜜凤梨
资源类型：选育品种
主要用途：鲜食或加工
系谱：开英种（♀）× Rough（♂）的杂交后代
选育单位：中国台湾农业试验所嘉义农业试验站
植株姿态：开张
定植期：2021 年 2 月中旬
现红期：2022 年 3 月上旬
营养生长期：约 380 d
初花期：2022 年 3 月下旬至 4 月上旬
花开放时间：约 15 d
成熟期：2022 年 6 月下旬至 7 月上旬
果实发育期：约 90 d
果实形状：长圆筒形
单果质量：1423.1 g
纵径：16.4 cm
横径：11.0 cm
果形指数：1.5
未成熟果实果皮颜色：银绿色
成熟果实果皮颜色：亮黄 / 淡黄色
果颈：有
果基：弧形
果顶：平
果实小果能否剥离：不可剥离
果眼外观：微隆起
果眼深度：较浅
果眼排列方式：右旋
果瘤：无
果肉颜色：淡黄色
果实香味：芳香
果实风味：浓甜
果肉质地：滑
果实外观综合评价：优
果实品质综合评价：优
果肉可溶性固形物含量：18.9%
果实成熟特性：晚熟
成熟期的一致性：基本一致
丰产性：高产

Tainung No. 16

Code: ACCESSION000020
Name: Tainung No. 16, also known as Tianmimi Pineapple
Resource type: bred variety
Main ways of consumption: fresh or processing
Lineage: a hybrid between Smooth Cayenne (♀) and Rough (♂)
Breeding organization: Chiayi Agricultural Experiment Station, Taiwan Agricultural Research Institute, China
Plant posture: spreading
Planting time: mid-February in 2021
Open heart stage: early March in 2022
Vegetative growth period: approx. 380 d
Initial flowering time: late March to early April in 2022
Flowering period: approx. 15 d
Fruit maturation time: late June to early July in 2022
Fruit developing period: approx. 90 d
Fruit shape: long cylindrical
Single fruit weight: 1423.1 g
Longitudinal diameter: 16.4 cm
Transverse diameter: 11.0 cm
Fruit shape index: 1.5
Immature fruit peel color: silver green
Mature fruit peel color: bright yellow/light yellow
Fruit neck: present
Fruit base: curved
Fruit top: flat
Fruitlets adhesion situation: inseparable
Fruit eyes appearance: slightly embossed
Fruit eyes depth: relatively shallow
Fruit eyes arrangement: dextrorotation
Fruit tumors: absent
Flesh color: pale yellow
Fruit aroma: sweet fragrance
Fruit flavour: strong sweet
Flesh texture: smooth
Fruit appearance comprehensive evaluation: excellent
Fruit quality comprehensive evaluation: excellent
Soluble solids content in flesh: 18.9%
Fruit ripening characteristics: late mature
Maturity consistency: basically uniform
Productivity: productive

植株
Plant

现红
Open heart

花
Flower

花序
Inflorescence

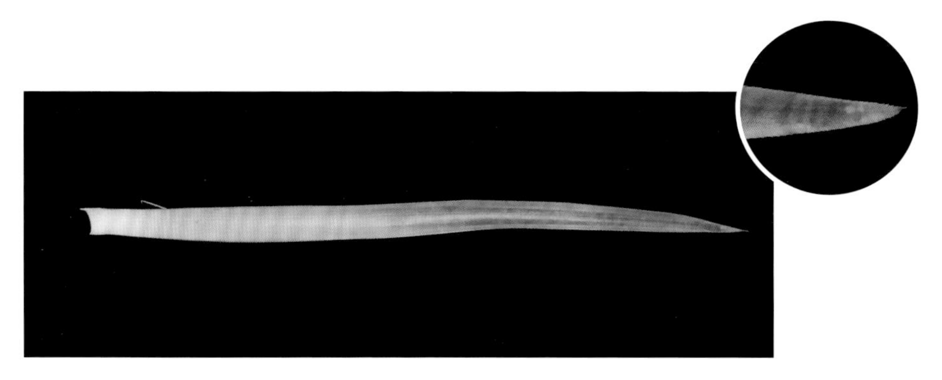

叶片
Leaf

带冠芽果
Fruit with crown bud

冠芽叶刺
Leaf spines in crown bud

果实
Fruit

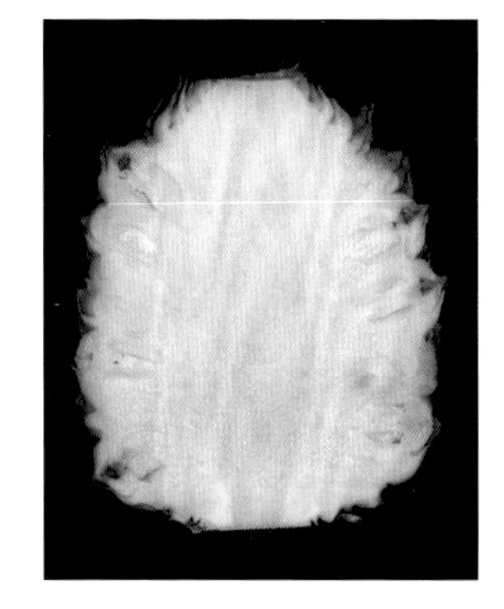

果实纵切
Fruit longitudinal section

果实横切
Fruit transverse section

台农 18 号

编号：ACCESSION000044
种质名称：台农 18 号，又名金桂花凤梨
资源类型：选育品种
主要用途：鲜食或加工
系谱：开英种（♀）× 1（A）1（♂）的杂交后代
选育单位：中国台湾农业试验所嘉义农业试验站
植株姿态：开张
定植期：2021 年 2 月中旬
现红期：2022 年 2 月下旬至 3 月上旬
营养生长期：约 380 d
初花期：2022 年 3 月下旬至 4 月上旬
花开放时间：约 25 d
成熟期：2022 年 6 月下旬至 7 月上旬
果实发育期：约 90 d
果实形状：圆筒形
单果质量：788.1 g
纵径：13.1 cm
横径：10.2 cm
果形指数：1.3
未成熟果实果皮颜色：淡绿色
成熟果实果皮颜色：亮黄色
果颈：无
果基：突起
果顶：平
果实小果能否剥离：不可剥离
果眼外观：微隆起
果眼深度：浅
果眼排列方式：右旋
果瘤：无
果肉颜色：淡黄色
果实香味：清香 / 微香
果实风味：浓甜
果肉质地：滑
果实外观综合评价：中
果实品质综合评价：好
果肉可溶性固形物含量：18.6%
果实成熟特性：中熟
成熟期的一致性：基本一致
丰产性：低产

Tainung No. 18

Code: ACCESSION000044
Name: Tainung No. 18, also known as Jinguihua Pineapple
Resource type: bred variety
Main ways of consumption: fresh or processing
Lineage: a hybrid between Smooth Cayenne (♀) and 1(A)1 (♂)
Breeding organization: Chiayi Agricultural Experiment Station, Taiwan Agricultural Research Institute, China
Plant posture: spreading
Planting time: mid-February in 2021
Open heart stage: late February to early March in 2022
Vegetative growth period: approx. 380 d
Initial flowering time: late March to early April in 2022
Flowering period: approx. 25 d
Fruit maturation time: late June to early July in 2022
Fruit developing period: approx. 90 d
Fruit shape: cylindrical
Single fruit weight: 788.1 g
Longitudinal diameter: 13.1 cm
Transverse diameter: 10.2 cm
Fruit shape index: 1.3
Immature fruit peel color: light green
Mature fruit peel color: bright yellow
Fruit neck: absent
Fruit base: protrusive
Fruit top: flat
Fruitlets adhesion situation: inseparable
Fruit eyes appearance: slightly embossed
Fruit eyes depth: shallow
Fruit eyes arrangement: dextrorotation
Fruit tumors: absent
Flesh color: pale yellow
Fruit aroma: faint/slight scent
Fruit flavour: strong sweet
Flesh texture: smooth
Fruit appearance comprehensive evaluation: medium
Fruit quality comprehensive evaluation: good
Soluble solids content in flesh: 18.6%
Fruit ripening characteristics: medium mature
Maturity consistency: basically uniform
Productivity: low

植株
Plant

现红
Open heart

花
Flower

花序
Inflorescence

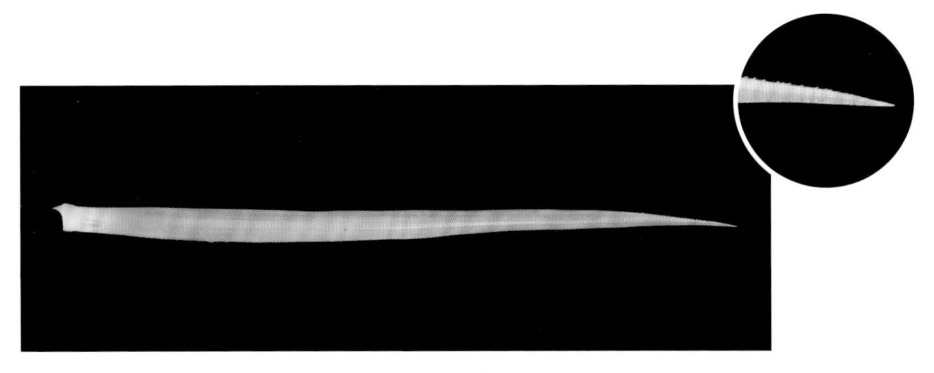

叶片
Leaf

带冠芽果
Fruit with crown bud

冠芽叶刺
Leaf spines in crown bud

果实
Fruit

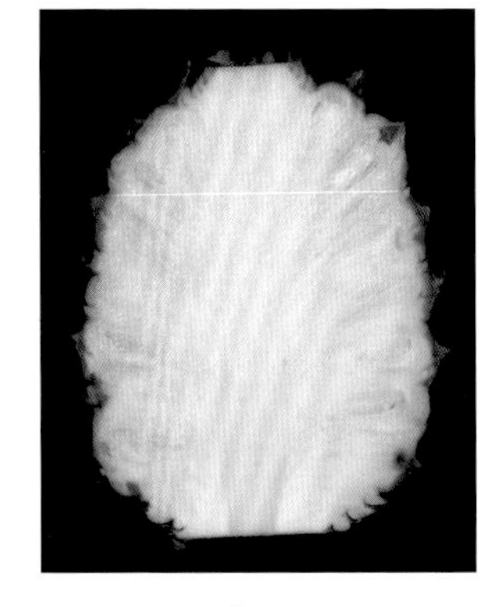

果实纵切
Fruit longitudinal section

果实横切
Fruit transverse section

台农 20 号

编号：ACCESSION000005
种质名称：台农 20 号，又名牛奶凤梨
资源类型：选育品种
主要用途：鲜食或加工
系谱：芽条变异选育
选育单位：中国台湾农业试验所凤山热带园艺试验分所
植株姿态：直立
定植期：2021 年 12 月下旬
现红期：2023 年 3 月上中旬
营养生长期：约 435 d
初花期：2023 年 4 月上旬
花开放时间：约 16 d
成熟期：2023 年 8 月上旬
果实发育期：约 120 d
果实形状：椭圆形
单果质量：488.8 g
纵径：9.9 cm
横径：9.3 cm
果形指数：1.1
未成熟果实果皮颜色：暗墨绿色
成熟果实果皮颜色：黄色，带绿斑
果颈：有
果基：平
果顶：钝圆
果实小果能否剥离：不可剥离
果眼外观：微隆起
果眼深度：浅
果眼排列方式：左旋
果瘤：无
果肉颜色：奶油色
果实香味：芳香
果实风味：酸甜
果肉质地：滑
果实外观综合评价：中
果实品质综合评价：中
果肉可溶性固形物含量：9.3%
果实成熟特性：晚熟
成熟期的一致性：不一致
丰产性：低产

Tainung No. 20

Code: ACCESSION000005
Name: Tainung No. 20, also known as Milk Pineapple
Resource type: bred variety
Main ways of consumption: fresh or processing
Lineage: bud mutation breeding
Breeding organization: Fengshan Tropical Horticulture Experimental Branch, Taiwan Agricultural Research Institute, China
Plant posture: upright
Planting time: late December in 2021
Open heart stage: early to mid-March in 2023
Vegetative growth period: approx. 435 d
Initial flowering time: early April in 2023
Flowering period: approx. 16 d
Fruit maturation time: early August in 2023
Fruit developing period: approx. 120 d
Fruit shape: ellipsoid
Single fruit weight: 488.8 g
Longitudinal diameter: 9.9 cm
Transverse diameter: 9.3 cm
Fruit shape index: 1.1
Immature fruit peel color: dark blackish green
Mature fruit peel color: yellow with green stripe
Fruit neck: present
Fruit base: flat
Fruit top: bluntly round
Fruitlets adhesion situation: inseparable
Fruit eyes appearance: slightly embossed
Fruit eyes depth: shallow
Fruit eyes arrangement: levorotation
Fruit tumors: absent
Flesh color: cream
Fruit aroma: sweet fragrance
Fruit flavour: sour-sweet
Flesh texture: smooth
Fruit appearance comprehensive evaluation: medium
Fruit quality comprehensive evaluation: medium
Soluble solids content in flesh: 9.3%
Fruit ripening characteristics: late mature
Maturity consistency: nonuniform
Productivity: low

植株
Plant

现红
Open heart

花
Flower

花序
Inflorescence

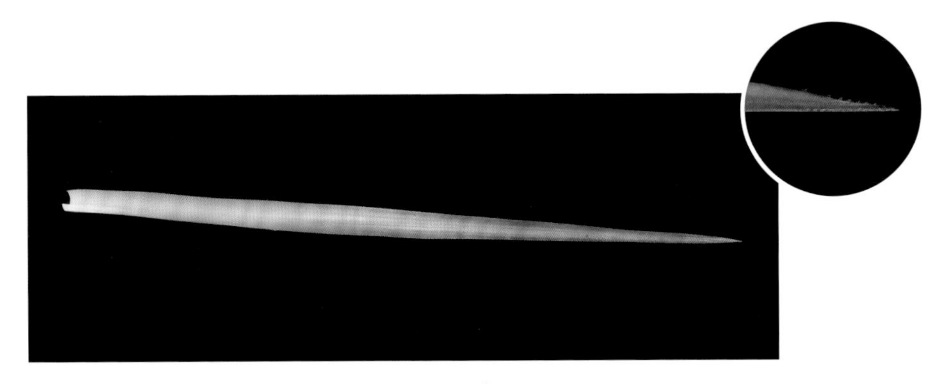

叶片
Leaf

带冠芽果
Fruit with crown bud

冠芽叶刺
Leaf spines in crown bud

果实
Fruit

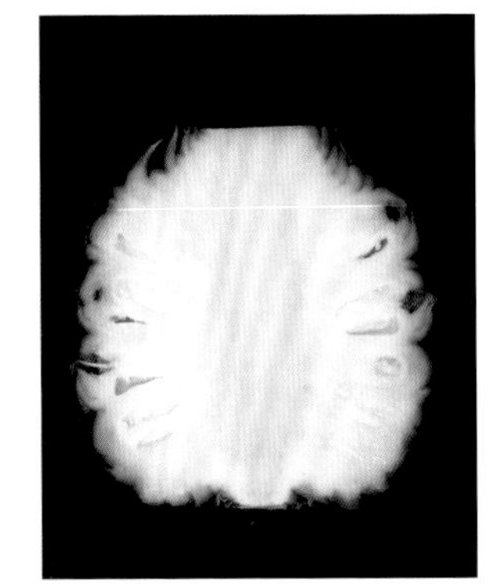

果实纵切
Fruit longitudinal section

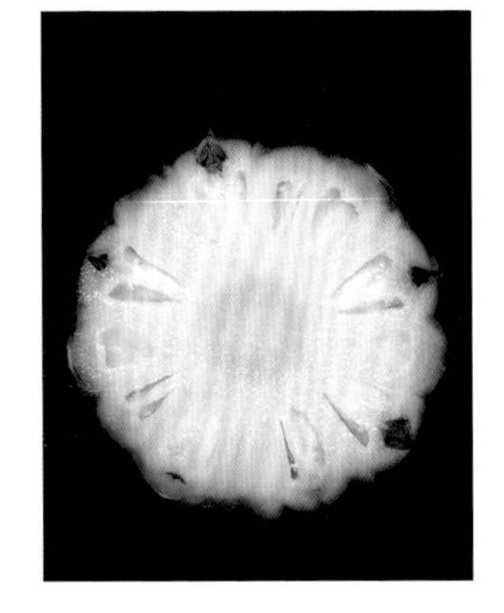

果实横切
Fruit transverse section

西瓜凤梨

编号：ACCESSION000104
种质名称：西瓜凤梨
原产地：不详
资源类型：国内收集
主要用途：鲜食或加工
种质来源地：2015 年中国热带农业科学院南亚热带作物研究所从海南收集
植株姿态：开张
定植期：2022 年 1 月上旬
现红期：2023 年 3 月中旬
营养生长期：约 440 d
初花期：2023 年 4 月上旬
花开放时间：约 16 d
成熟期：2023 年 7 月中旬
果实发育期：约 100 d
果实形状：近圆球形
单果质量：1728.5 g
纵径：14.6 cm
横径：13.8 cm
果形指数：1.1
未成熟果实果皮颜色：暗墨绿色
成熟果实果皮颜色：暗黄色
果颈：无
果基：平
果顶：浑圆
果实小果能否剥离：可剥离
果眼外观：微隆起
果眼深度：浅
果眼排列方式：左旋
果瘤：无
果肉颜色：黄色 / 金黄色
果实香味：清香 / 微香
果实风味：清甜至甜酸
果肉质地：脆 / 爽脆
果实外观综合评价：好
果实品质综合评价：中
果肉可溶性固形物含量：15.2%
果实成熟特性：晚熟
成熟期的一致性：基本一致
丰产性：丰产

Watermelon Pineapple

Code: ACCESSION000104
Name: Watermelon Pineapple
Place of origin: unknown
Resource type: domestically collected variety
Main ways of consumption: fresh or processing
Material resource: SSCRI, CATAS collected from Hainan Province in 2015
Plant posture: spreading
Planting time: early January in 2022
Open heart stage: mid-March in 2023
Vegetative growth period: approx. 440 d
Initial flowering time: early April in 2023
Flowering period: approx. 16 d
Fruit maturation time: mid-July in 2023
Fruit developing period: approx.100 d
Fruit shape: near spherical
Single fruit weight: 1728.5 g
Longitudinal diameter: 14.6 cm
Transverse diameter: 13.8 cm
Fruit shape index: 1.1
Immature fruit peel color: dark blackish green
Mature fruit peel color: dark yellow
Fruit neck: absent
Fruit base: flat
Fruit top: perfectly round
Fruitlets adhesion situation: separable
Fruit eyes appearance: slightly embossed
Fruit eyes depth: shallow
Fruit eyes arrangement: levorotation
Fruit tumors: absent
Flesh color: yellow/golden yellow
Fruit aroma: faint/slight scent
Fruit flavour: mildly sweet to sweet-sour
Flesh texture: crisp/crunchy
Fruit appearance comprehensive evaluation: good
Fruit quality comprehensive evaluation: medium
Soluble solids content in flesh: 15.2%
Fruit ripening characteristics: late mature
Maturity consistency: basically uniform
Productivity: productive

植株
Plant

现红
Open heart

花
Flower

花序
Inflorescence

叶片
Leaf

带冠芽果
Fruit with crown bud

冠芽叶刺
Leaf spines in crown bud

果实
Fruit

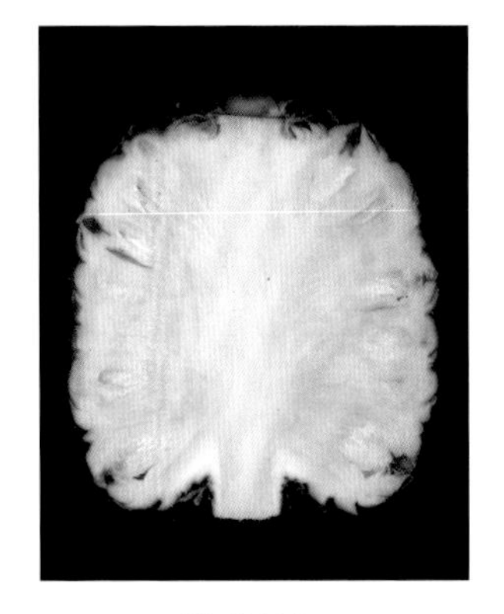

果实纵切
Fruit longitudinal section

果实横切
Fruit transverse section

台农 23 号

编号：ACCESSION000117
种质名称：台农 23 号，又名芒果凤梨
资源类型：选育品种
主要用途：鲜食或加工
系谱：台农 19 号（♀）× 台农 21 号（♂）的杂交后代
选育单位：中国台湾农业试验所嘉义农业试验分所
植株姿态：开张
定植期：2022 年 1 月中旬
现红期：2023 年 2 月下旬
营养生长期：约 410 d
初花期：2023 年 3 月下旬
花开放时间：约 16 d
成熟期：2023 年 6 月中旬至 7 月上旬
果实发育期：约 90 d
果实形状：短圆筒形
单果质量：1120.3 g
纵径：12.1 cm
横径：12.4 cm
果形指数：1.0
未成熟果实果皮颜色：暗绿色
成熟果实果皮颜色：橙红色
果颈：无
果基：平
果顶：浑圆
果实小果能否剥离：可剥离
果眼外观：微隆起
果眼深度：浅
果眼排列方式：左旋
果瘤：无或多
果肉颜色：淡黄色
果实香味：清香 / 微香
果实风味：酸甜至香甜
果肉质地：滑、脆 / 爽脆
果实外观综合评价：优
果实品质综合评价：好
果肉可溶性固形物含量：20.8%
果实成熟特性：中熟
成熟期的一致性：不一致
丰产性：丰产

Tainung No. 23

Code: ACCESSION000117
Name: Tainung No. 23, also known as Mango Pineapple
Resource type: bred variety
Main ways of consumption: fresh or processing
Lineage: a hybrid between Tainung No. 19 (♀) and Tainung No. 21 (♂)
Breeding organization: Chiayi Agricultural Experiment Experiment Branch, Taiwan Agricultural Research Institute, China
Plant posture: spreading
Planting time: early January in 2022
Open heart stage: late February in 2022
Vegetative growth period: approx. 410 d
Initial flowering time: late March in 2023
Flowering period: approx. 16 d
Fruit maturation time: mid-June to early July in 2023
Fruit developing period: approx. 90 d
Fruit shape: short cylindrical
Single fruit weight: 1120.3 g
Longitudinal diameter: 12.1 cm
Transverse diameter: 12.4 cm
Fruit shape index: 1.0
Immature fruit peel color: dark green
Mature fruit peel color: orange red
Fruit neck: absent
Fruit base: flat
Fruit top: perfectly round
Fruitlets adhesion situation: separable
Fruit eyes appearance: slightly embossed
Fruit eyes depth: shallow
Fruit eyes arrangement: levorotation
Fruit tumors: absent or many
Flesh color: light yellow
Fruit aroma: faint/slight scent
Fruit flavour: sour-sweet to fragrant and sweet
Flesh texture: smooth, crisp/crunchy
Fruit appearance comprehensive evaluation: excellent
Fruit quality comprehensive evaluation: good
Soluble solids content in flesh: 20.8%
Fruit ripening characteristics: medium mature
Maturity consistency: nonuniform
Productivity: productive

植株
Plant

现红
Open heart

花
Flower

花序
Inflorescence

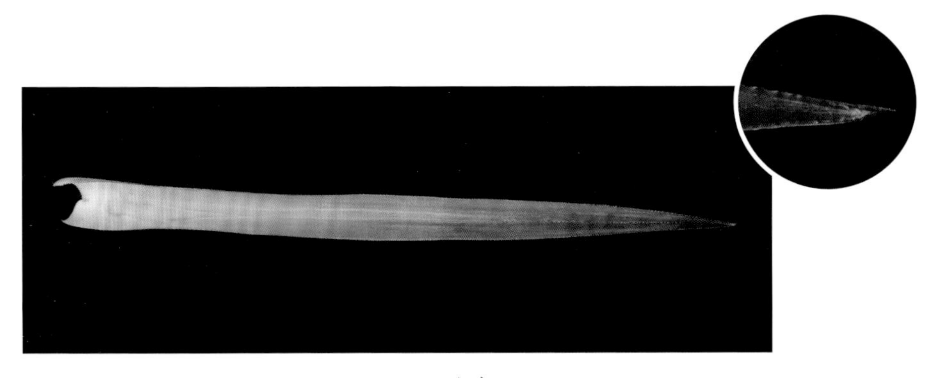

叶片
Leaf

带冠芽果
Fruit with crown bud

冠芽叶刺
Leaf spines in crown bud

果实
Fruit

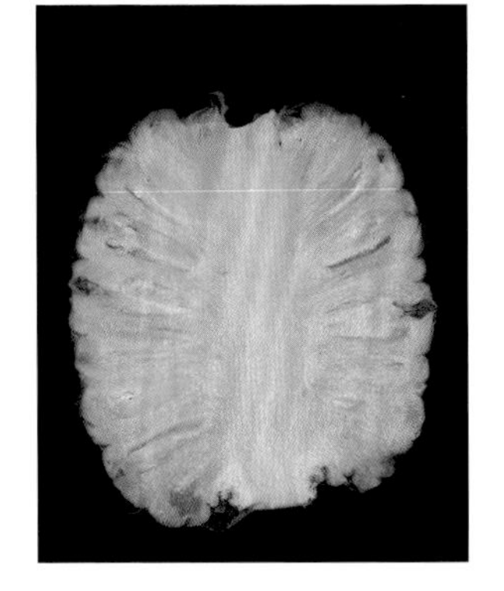

果实纵切
Fruit longitudinal section

果实横切
Fruit transverse section

台农 11 号多倍体

编号：ACCESSION000065
种质名称：台农 11 号多倍体
资源类型：国内收集
主要用途：鲜食或加工
种质来源地：2011 年中国热带农业科学院南亚热带作物研究所从广东省收集
植株姿态：开张
定植期：2021 年 12 月下旬
现红期：2023 年 3 月上旬
营养生长期：约 435 d
初花期：2023 年 4 月上旬
花开放时间：约 18 d
成熟期：2023 年 6 月上旬
果实发育期：约 60 d
果实形状：长圆筒形
单果质量：574.9 g
纵径：11.5 cm
横径：9.2 cm
果形指数：1.2
未成熟果实果皮颜色：暗墨绿色
成熟果实果皮颜色：其他
果颈：无
果基：平
果顶：钝圆
果实小果能否剥离：可剥离
果眼外观：突起 / 隆起
果眼深度：浅
果眼排列方式：左旋
果瘤：多
果肉颜色：淡黄色
果实香味：清香 / 微香
果实风味：酸甜
果肉质地：脆 / 爽脆
果实外观综合评价：中
果实品质综合评价：差
果肉可溶性固形物含量：14.6%
果实成熟特性：早熟
成熟期的一致性：不一致
丰产性：低产

Tainung No. 11 Polyploid

Code: ACCESSION000065
Name: Tainung No. 11 Polyploid
Resource type: domestically collected variety
Main ways of consumption: fresh or processing
Material resource: SSCRI, CATAS collected from Guangdong Province in 2011
Plant posture: spreading
Planting time: late December in 2021
Open heart stage: early March in 2023
Vegetative growth period: approx. 435 d
Initial flowering time: early April in 2023
Flowering period: approx. 18 d
Fruit maturation time: early June in 2023
Fruit developing period: approx. 60 d
Fruit shape: long cylindrical
Single fruit weight: 574.9 g
Longitudinal diameter: 11.5 cm
Transverse diameter: 9.2 cm
Fruit shape index: 1.2
Immature fruit peel color: dark blackish green
Mature fruit peel color: other
Fruit neck: absent
Fruit base: flat
Fruit top: bluntly round
Fruitlets adhesion situation: separable
Fruit eyes appearance: convex/embossed
Fruit eyes depth: shallow
Fruit eyes arrangement: levorotation
Fruit tumors: many
Flesh color: pale yellow
Fruit aroma: faint/slight scent
Fruit flavour: sour-sweet
Flesh texture: crisp/crunchy
Fruit appearance comprehensive evaluation: medium
Fruit quality comprehensive evaluation: poor
Soluble solids content in flesh: 14.6%
Fruit ripening characteristics: early mature
Maturity consistency: nonuniform
Productivity: low

植株
Plant

现红
Open heart

花
Flower

花序
Inflorescence

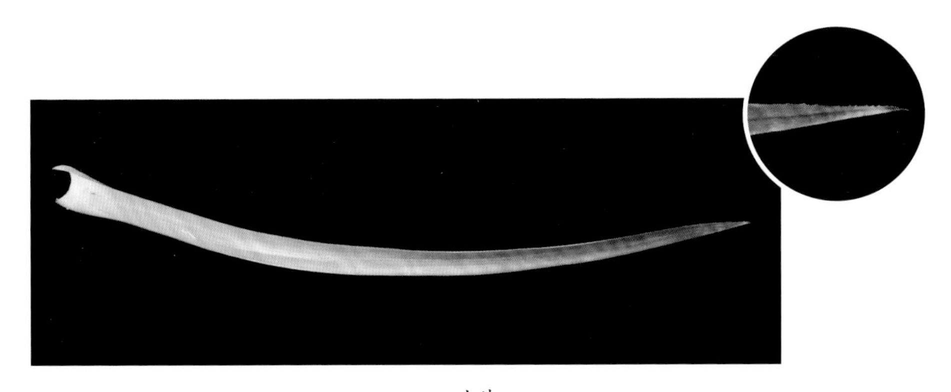

叶片
Leaf

带冠芽果
Fruit with crown bud

冠芽叶刺
Leaf spines in crown bud

果实
Fruit

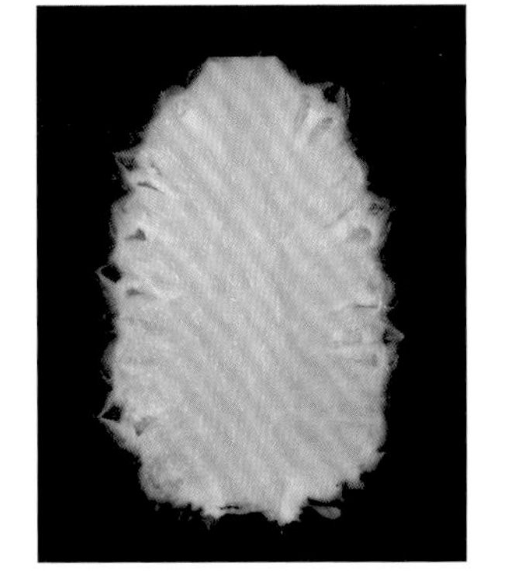

果实纵切
Fruit longitudinal section

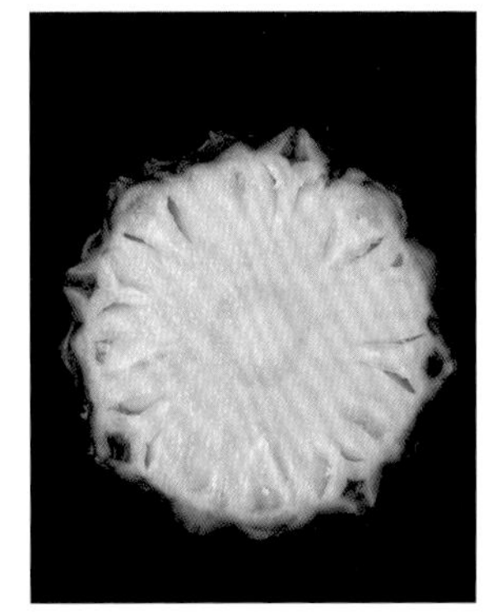

果实横切
Fruit transverse section

台湾有刺

编号：ACCESSION000067
种质名称：台湾有刺
原产地：不详
资源类型：国内收集
主要用途：鲜食或加工
种质来源地：2009年中国热带农业科学院南亚热带作物研究所从福建省漳州市收集
植株姿态：开张
定植期：2021年12月下旬
现红期：2023年2月下旬
营养生长期：约425 d
初花期：2023年3月下旬
花开放时间：约20 d
成熟期：2023年6月中旬
果实发育期：约80 d
果实形状：圆筒形
单果质量：680.9 g
纵径：12.1 cm
横径：10.1 cm
果形指数：1.2
未成熟果实果皮颜色：银绿色
成熟果实果皮颜色：亮黄色
果颈：无
果基：平
果顶：平
果实小果能否剥离：不可剥离
果眼外观：突起 / 隆起
果眼深度：深
果眼排列方式：右旋
果瘤：无
果肉颜色：黄色
果实香味：芳香
果实风味：甜酸
果肉质地：滑
果实外观综合评价：优
果实品质综合评价：好
果肉可溶性固形物含量：17.9%
果实成熟特性：中熟
成熟期的一致性：一致
丰产性：低产

Taiwan Youci

Code: ACCESSION000067
Name: Taiwan Youci
Place of origin: unknown
Resource type: domestically collected variety
Main ways of consumption: fresh or processing
Material resource: SSCRI, CATAS collected from Zhangzhou City, Fujian Province in 2009
Plant posture: spreading
Planting time: late December in 2021
Open heart stage: late February in 2023
Vegetative growth period: approx. 425 d
Initial flowering time: late March in 2023
Flowering period: approx. 20 d
Fruit maturation time: mid-June in 2023
Fruit developing period: approx. 80 d
Fruit shape: cylindrical
Single fruit weight: 680.9 g
Longitudinal diameter: 12.1 cm
Transverse diameter: 10.1 cm
Fruit shape index: 1.2
Immature fruit peel color: silver green
Mature fruit peel color: bright yellow
Fruit neck: absent
Fruit base: flat
Fruit top: flat
Fruitlets adhesion situation: inseparable
Fruit eyes appearance: convex/embossed
Fruit eyes depth: deep
Fruit eyes arrangement: dextrorotation
Fruit tumors: absent
Flesh color: yellow
Fruit aroma: sweet fragrance
Fruit flavour: sweet-sour
Flesh texture: smooth
Fruit appearance comprehensive evaluation: excellent
Fruit quality comprehensive evaluation: good
Soluble solids content in flesh: 17.9%
Fruit ripening characteristics: medium mature
Maturity consistency: uniform
Productivity: low

植株
Plant

现红
Open heart

花
Flower

花序
Inflorescence

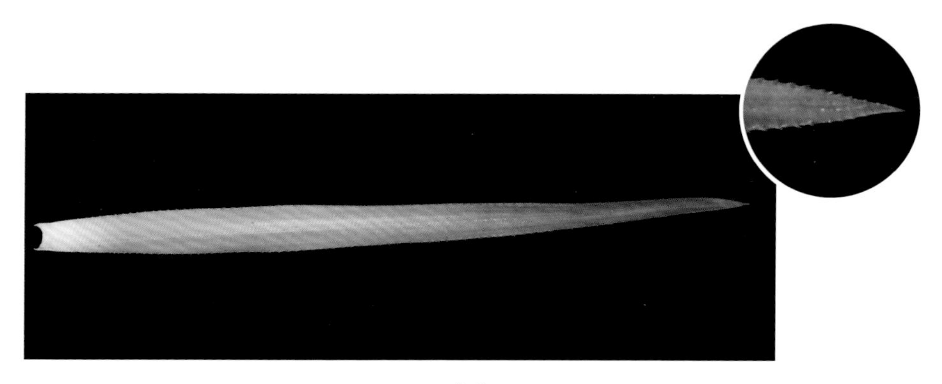

叶片
Leaf

带冠芽果
Fruit with crown bud

冠芽叶刺
Leaf spines in crown bud

果实
Fruit

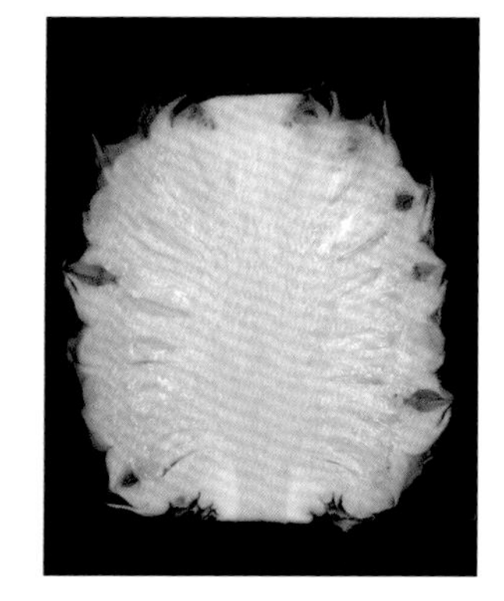

果实纵切
Fruit longitudinal section

果实横切
Fruit transverse section

粤脆

编号：ACCESSION000060
种质名称：粤脆
资源类型：选育品种
主要用途：鲜食或加工
系谱：无刺卡因 × 神湾的杂交后代
选育单位：广东省农业科学院果树研究所
植株姿态：开张
定植期：2021 年 12 月下旬
现红期：2023 年 2 月下旬至 3 月上旬
营养生长期：约 430 d
初花期：2023 年 3 月下旬至 4 月上旬
花开放时间：约 29 d
成熟期：2023 年 7 月下旬
果实发育期：约 115 d
果实形状：圆锥形
单果质量：798.7 g
纵径：11.3 cm
横径：11.3 cm
果形指数：1.0
未成熟果实果皮颜色：暗墨绿色
成熟果实果皮颜色：暗黄 / 深黄色
果颈：有
果基：平
果顶：钝圆
果实小果能否剥离：不可剥离
果眼外观：突起 / 隆起
果眼深度：浅
果眼排列方式：左旋或右旋
果瘤：无
果肉颜色：淡黄色
果实香味：无
果实风味：清甜
果肉质地：滑
果实外观综合评价：差
果实品质综合评价：中
果肉可溶性固形物含量：17.7%
果实成熟特性：晚熟
成熟期的一致性：基本一致
丰产性：低产

Yuecui

Code: ACCESSION000060
Name: Yuecui
Resource type: bred variety
Main ways of consumption: fresh or processing
Lineage: a hybrid between Smooth Cayenne and Shenwan
Breeding organization: Institute of Fruit Tree Research, Guangdong Academy of Agricultural Science
Plant posture: spreading
Planting time: late December in 2021
Open heart stage: late February to early March in 2023
Vegetative growth period: approx. 430 d
Initial flowering time: late March to early April in 2023
Flowering period: approx. 29 d
Fruit maturation time: late July in 2023
Fruit developing period: approx. 115 d
Fruit shape: conical
Single fruit weight: 798.7 g
Longitudinal diameter: 11.3 cm
Transverse diameter: 11.3 cm
Fruit shape index: 1.0
Immature fruit peel color: dark blackish green
Mature fruit peel color: dark yellow/deep yellow
Fruit neck: present
Fruit base: flat
Fruit top: bluntly round
Fruitlets adhesion situation: inseparable
Fruit eyes appearance: convex/embossed
Fruit eyes depth: shallow
Fruit eyes arrangement: levorotation or dextrorotation
Fruit tumors: absent
Flesh color: light yellow
Fruit aroma: no scent
Fruit flavour: mildly sweet
Flesh texture: smooth
Fruit appearance comprehensive evaluation: poor
Fruit quality comprehensive evaluation: medium
Soluble solids content in flesh: 17.7%
Fruit ripening characteristics: late mature
Maturity consistency: basically uniform
Productivity: low

植株
Plant

现红
Open heart

花
Flower

花序
Inflorescence

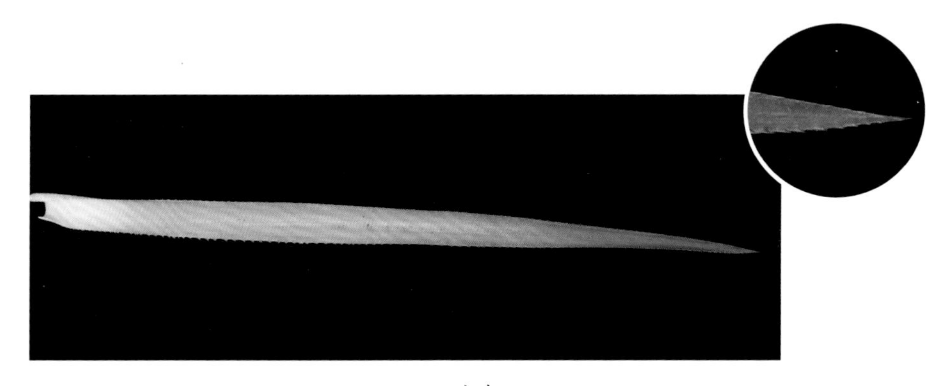

叶片
Lcaf

带冠芽果
Fruit with crown bud

冠芽叶刺
Leaf spines in crown bud

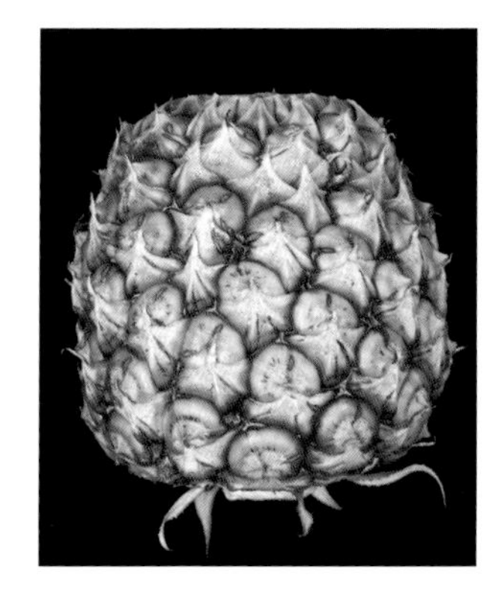

果实
Fruit

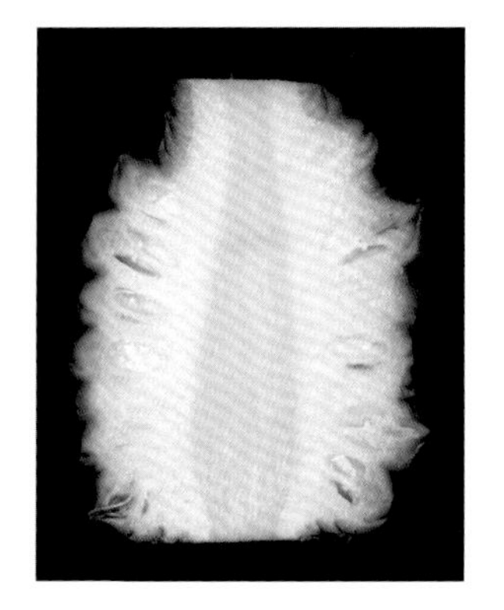

果实纵切
Fruit longitudinal section

果实横切
Fruit transverse section

云南引卡因

编号：ACCESSION000054
种质名称：云南引卡因
原产地：不详
资源类型：国内收集
主要用途：鲜食或加工
种质来源地：2008 年中国热带农业科学院南亚热带作物研究所从云南西双版纳傣族自治州收集
植株姿态：开张
定植期：2022 年 1 月上旬
现红期：2023 年 3 月上中旬
营养生长期：约 430 d
初花期：2023 年 4 月上旬
花开放时间：约 21 d
成熟期：2023 年 7 月中下旬
果实发育期：约 105 d
果实形状：圆筒形
单果质量：680.0 g
纵径：10.9 cm
横径：10.1 cm
果形指数：1.1
未成熟果实果皮颜色：暗绿色
成熟果实果皮颜色：暗黄 / 深黄色
果颈：无
果基：平
果顶：平
果实小果能否剥离：不可剥离
果眼外观：微隆起
果眼深度：浅
果眼排列方式：左旋
果瘤：无
果肉颜色：淡黄色
果实香味：清香 / 微香
果实风味：酸甜
果肉质地：粗糙
果实外观综合评价：优
果实品质综合评价：中
果肉可溶性固形物含量：13.5%
果实成熟特性：晚熟
成熟期的一致性：不一致
丰产性：低产

Yunnan Smooth Cayenne

Code: ACCESSION000054
Name: Yunnan Smooth Cayenne
Place of origin: unknown
Resource type: collected variety
Main ways of consumption: fresh or processing
Material resource: SSCRI, CATAS collected from Xishuangbanna Dai Autonomous Prefecture, Yunnan Province in 2008
Plant posture: spreading
Planting time: early Juanary in 2022
Open heart stage: early to mid-March in 2023
Vegetative growth period: approx. 430 d
Initial flowering time: early April in 2023
Flowering period: approx. 21 d
Fruit maturation time: mid to late July in 2023
Fruit developing period: approx. 105 d
Fruit shape: cylindrical
Single fruit weight: 680.0 g
Longitudinal diameter: 10.9 cm
Transverse diameter: 10.1 cm
Fruit shape index: 1.1
Immature fruit peel color: dark green
Mature fruit peel color: dark yellow/deep yellow
Fruit neck: absent
Fruit base: flat
Fruit top: flat
Fruitlets adhesion situation: inseparable
Fruit eyes appearance: slightly embossed
Fruit eyes depth: shallow
Fruit eyes arrangement: levorotation
Fruit tumors: absent
Flesh color: light yellow
Fruit aroma: faint/slight scent
Fruit flavour: sour-sweet
Flesh texture: coarse
Fruit appearance comprehensive evaluation: excellent
Fruit quality comprehensive evaluation: medium
Soluble solids content in flesh: 13.5%
Fruit ripening characteristics: late mature
Maturity consistency: nonuniform
Productivity: low

植株
Plant

现红
Open heart

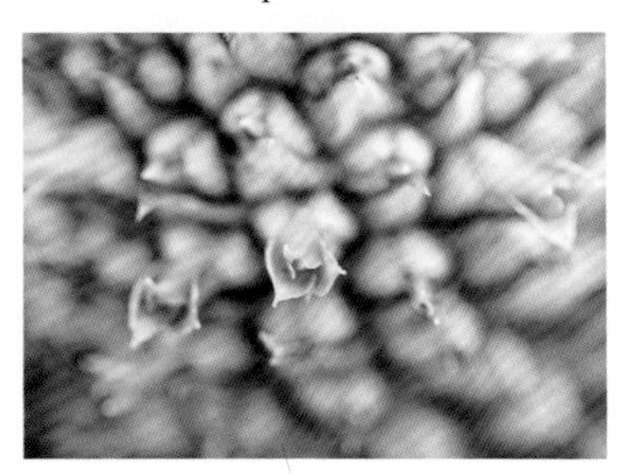

花
Flower

花序
Inflorescence

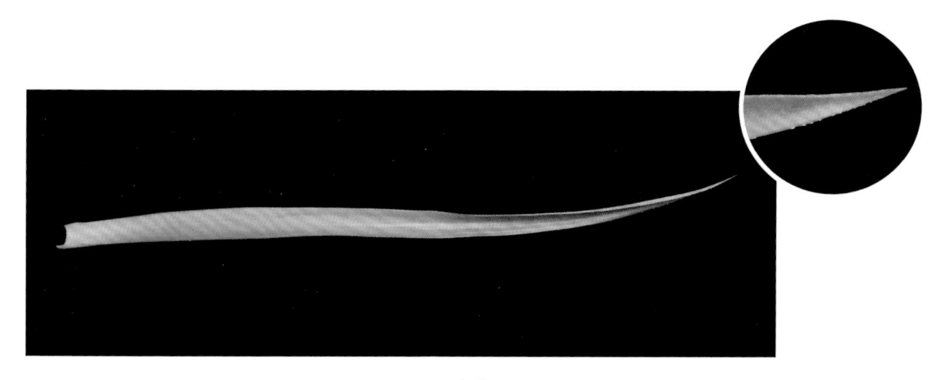

叶片
Lcaf

带冠芽果
Fruit with crown bud

冠芽叶刺
Leaf spines in crown bud

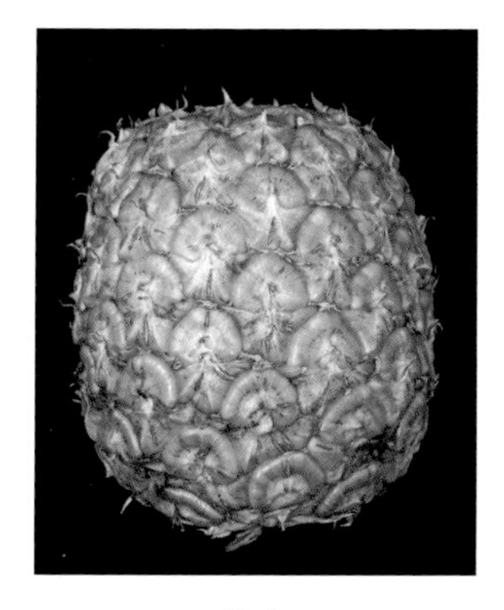

果实
Fruit

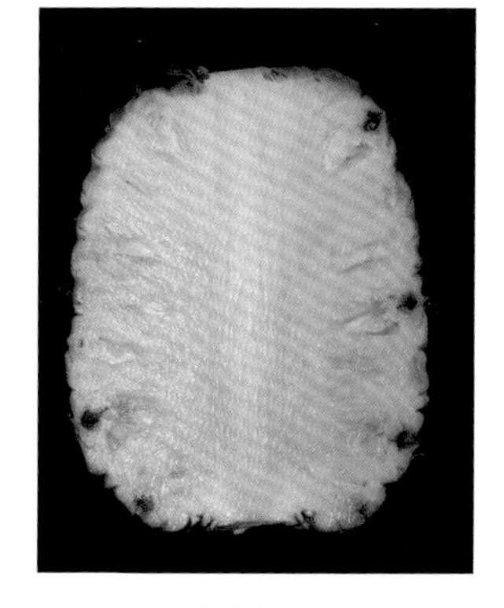

果实纵切
Fruit longitudinal section

果实横切
Fruit transverse section

云南版纳无刺卡因

编号：ACCESSION000105
种质名称：云南版纳无刺卡因
原产地：不详
资源类型：国内收集
主要用途：鲜食或加工
种质来源地：2016 年中国热带农业科学院南亚热带作物研究所从云南省西双版纳傣族自治州收集
植株姿态：开张
定植期：2022 年 1 月上旬
现红期：2023 年 3 月上旬
营养生长期：约 435 d
初花期：2023 年 4 月上旬
花开放时间：约 20 d
成熟期：2023 年 7 月中旬
果实发育期：约 100 d
果实形状：长圆筒形
单果质量：1482.4 g
纵径：15.3 cm
横径：12.0 cm
果形指数：1.3
未成熟果实果皮颜色：暗墨绿色
成熟果实果皮颜色：金黄色
果颈：无
果基：平
果顶：平
果实小果能否剥离：不可剥离
果眼外观：微隆起
果眼深度：浅
果眼排列方式：左旋
果瘤：无
果肉颜色：淡黄色
果实香味：清香 / 微香
果实风味：酸甜
果肉质地：滑
果实外观综合评价：好
果实品质综合评价：中
果肉可溶性固形物含量：17.1%
果实成熟特性：晚熟
成熟期的一致性：基本一致
丰产性：丰产

Yunnan Banna Smooth Cayenne

Code: ACCESSION000105
Name: Yunnan Banna Smooth Cayenne
Place of origin: unknown
Resource type: domestically collected variety
Main ways of consumption: fresh or processing
Material resource: SSCRI, CATAS collected from Xishuangbanna Dai Autonomous Prefecture, Yunnan Province in 2016
Plant posture: spreading
Planting time: early Juanary in 2022
Open heart stage: early March in 2023
Vegetative growth period: approx. 435 d
Initial flowering time: early April in 2023
Flowering period: approx. 20 d
Fruit maturation time: mid-July in 2023
Fruit developing period: approx. 100 d
Fruit shape: long cylindrical
Single fruit weight: 1482.4 g
Longitudinal diameter: 15.3 cm
Transverse diameter: 12.0 cm
Fruit shape index: 1.3
Immature fruit peel color: dark blackish green
Mature fruit peel color: golden yellow
Fruit neck: absent
Fruit base: flat
Fruit top: flat
Fruitlets adhesion situation: inseparable
Fruit eyes appearance: slightly embossed
Fruit eyes depth: shallow
Fruit eyes arrangement: levorotation
Fruit tumors: absent
Flesh color: light yellow
Fruit aroma: faint/slight scent
Fruit flavour: sour-sweet
Flesh texture: smooth
Fruit appearance comprehensive evaluation: good
Fruit quality comprehensive evaluation: medium
Soluble solids content in flesh: 17.1%
Fruit ripening characteristics: late mature
Maturity consistency: basically uniform
Productivity: productive

植株
Plant

现红
Open heart

花
Flower

花序
Inflorescence

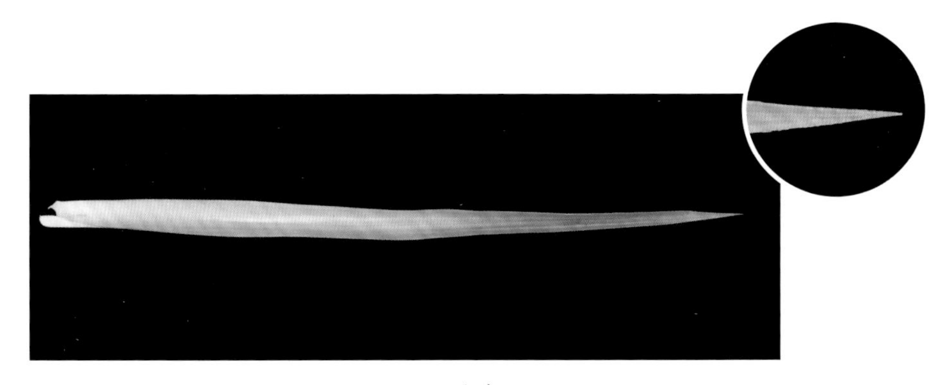

叶片
Leaf

带冠芽果
Fruit with crown bud

冠芽叶刺
Leaf spines in crown bud

果实
Fruit

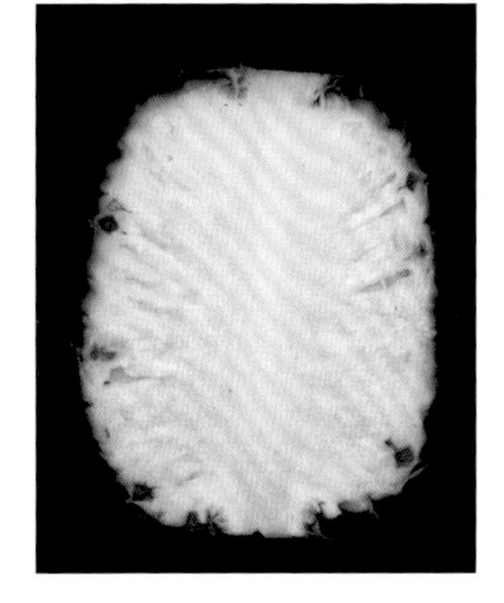

果实纵切
Fruit longitudinal section

果实横切
Fruit transverse section

云南引未知名越南种

编号：ACCESSION000037
种质名称：云南引未知名越南种
原产地：越南
资源类型：国内收集
主要用途：鲜食或加工
种质来源地：2012年中国热带农业科学院南亚热带作物研究所从云南省收集
植株姿态：开张
定植期：2021年2月下旬
现红期：2022年2月下旬至3月上旬
营养生长期：约370 d
初花期：2022年3月下旬至4月上旬
花开放时间：约17 d
成熟期：2022年6月中旬
果实发育期：约85 d
果实形状：圆筒形
单果质量：397.4 g
纵径：9.9 cm
横径：8.8 cm
果形指数：1.1
未成熟果实果皮颜色：暗绿色
成熟果实果皮颜色：深黄至橙色
果颈：无
果基：平
果顶：平
果实小果能否剥离：不可剥离
果眼外观：突起 / 隆起
果眼深度：较浅
果眼排列方式：左旋
果瘤：无
果肉颜色：淡黄色
果实香味：清香 / 微香
果实风味：甜酸
果肉质地：滑
果实外观综合评价：中
果实品质综合评价：好
果肉可溶性固形物含量：17.0%
果实成熟特性：中熟
成熟期的一致性：一致
丰产性：低产

Unknown Vietnam Cultivar from Yunnan

Code: ACCESSION000037
Name: Unknown Vietnam Cultivar from Yunnan
Place of origin: Vietnam
Resource type: domestically collected variety
Main ways of consumption: fresh or processing
Material resource: SSCRI, CATAS collected from Yunnan Province in 2012
Plant posture: spreading
Planting time: late February in 2021
Open heart stage: late February to early March in 2022
Vegetative growth period: approx. 370 d
Initial flowering time: late March to early April in 2022
Flowering period: approx. 17 d
Fruit maturation time: mid-June in 2022
Fruit developing period: approx. 85 d
Fruit shape: cylindrical
Single fruit weight: 397.4 g
Longitudinal diameter: 9.9 cm
Transverse diameter: 8.8 cm
Fruit shape index: 1.1
Immature fruit peel color: dark green
Mature fruit peel color: from deep yellow to orange
Fruit neck: absent
Fruit base: flat
Fruit top: flat
Fruitlets adhesion situation: inseparable
Fruit eyes appearance: convex/embossed
Fruit eyes depth: relatively shallow
Fruit eyes arrangement: levorotation
Fruit tumors: absent
Flesh color:light yellow
Fruit aroma: faint/slight scent
Fruit flavour: sweet-sour
Flesh texture: smooth
Fruit appearance comprehensive evaluation: medium
Fruit quality comprehensive evaluation: good
Soluble solids content in flesh: 17.0%
Fruit ripening characteristics: medium mature
Maturity consistency: uniform
Productivity: low

植株
Plant

现红
Open heart

花
Flower

花序
Inflorescence

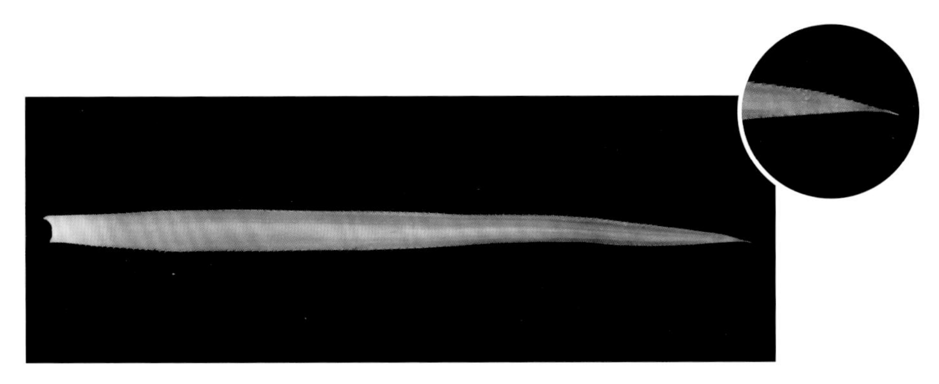

叶片
Leaf

带冠芽果
Fruit with crown bud

冠芽叶刺
Leaf spines in crown bud

果实
Fruit

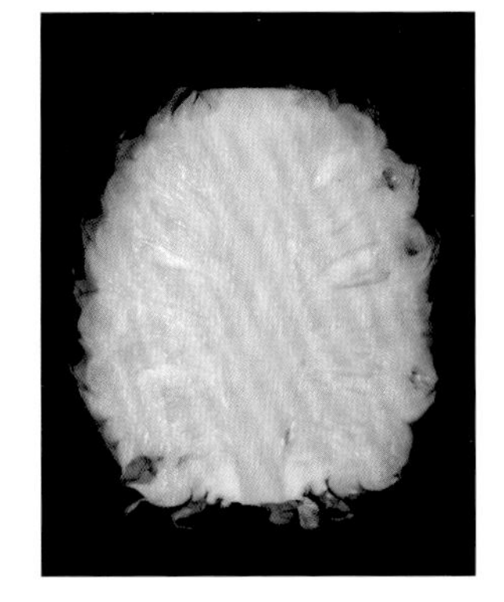

果实纵切
Fruit longitudinal section

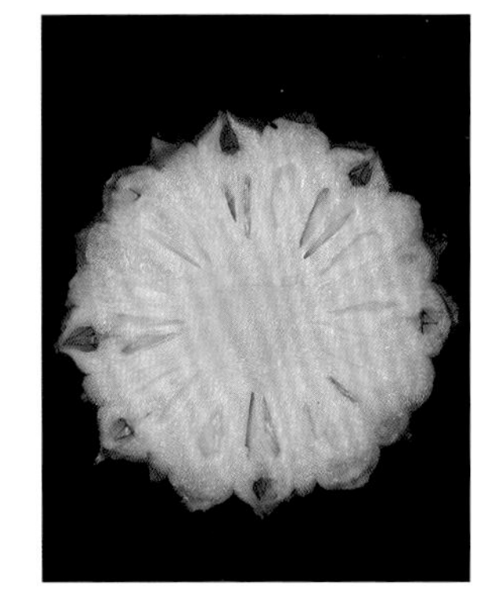

果实横切
Fruit transverse section

广西引沙捞越

编号：ACCESSION000068
种质名称：广西引沙捞越
原产地：马来西亚
资源类型：国内收集
主要用途：鲜食或加工
种质来源地：2009年中国热带农业科学院南亚热带作物研究所从广西壮族自治区收集
植株姿态：开张
定植期：2022年1月上旬
现红期：2023年3月上旬
营养生长期：约425 d
初花期：2023年4月上旬
花开放时间：约23 d
成熟期：2023年7月中下旬
果实发育期：约105 d
果实形状：圆筒形
单果质量：817.2 g
纵径：11.7 cm
横径：10.7 cm
果形指数：1.1
未成熟果实果皮颜色：暗绿色
成熟果实果皮颜色：暗黄色
果颈：无
果基：平
果顶：平
果实小果能否剥离：不可剥离
果眼外观：扁平或微凹
果眼深度：较浅
果眼排列方式：左旋或右旋
果瘤：无
果肉颜色：淡黄色
果实香味：清香 / 微香
果实风味：酸甜
果肉质地：粗糙
果实外观综合评价：中
果实品质综合评价：中
果肉可溶性固形物含量：15.6%
果实成熟特性：晚熟
成熟期的一致性：不一致
丰产性：低产

Sarawak from Guangxi

Code: ACCESSION000068
Name: Sarawak from Guangxi
Place of origin: Malaysia
Resource type: domestically collected variety
Main ways of consumption: fresh or processing
Material resource: SSCRI, CATAS collected from Guangxi Zhuang Autonomous Region in 2009
Plant posture: spreading
Planting time: early January in 2022
Open heart stage: early March in 2023
Vegetative growth period: approx. 425 d
Initial flowering time: early April in 2023
Flowering period: approx. 23 d
Fruit maturation time: mid to late July in 2023
Fruit developing period: approx. 105 d
Fruit shape: cylindrical
Single fruit weight: 817.2 g
Longitudinal diameter: 11.7 cm
Transverse diameter: 10.7 cm
Fruit shape index: 1.1
Immature fruit peel color: dark green
Mature fruit peel color: dark yellow
Fruit neck: absent
Fruit base: flat
Fruit top: flat
Fruitlets adhesion situation: inseparable
Fruit eyes appearance: flat or slightly concave
Fruit eyes depth: relatively shallow
Fruit eyes arrangement: levorotation or dextrorotation
Fruit tumors: absent
Flesh color: light yellow
Fruit aroma: faint/slight scent
Fruit flavour: sour-sweet
Flesh texture: coarse
Fruit appearance comprehensive evaluation: medium
Fruit quality comprehensive evaluation: medium
Soluble solids content in flesh: 15.6%
Fruit ripening characteristics: medium mature
Maturity consistency: nonuniform
Productivity: low

植株
Plant

现红
Open heart

花
Flower

花序
Inflorescence

叶片
Leaf

带冠芽果
Fruit with crown bud

冠芽叶刺
Leaf spines in crown bud

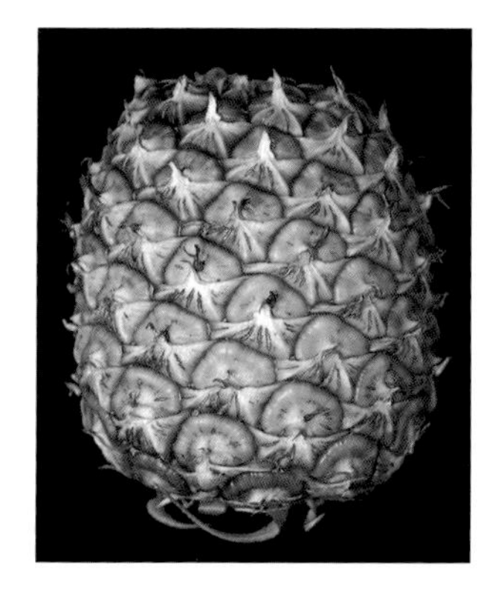

果实
Fruit

果实纵切
Fruit longitudinal section

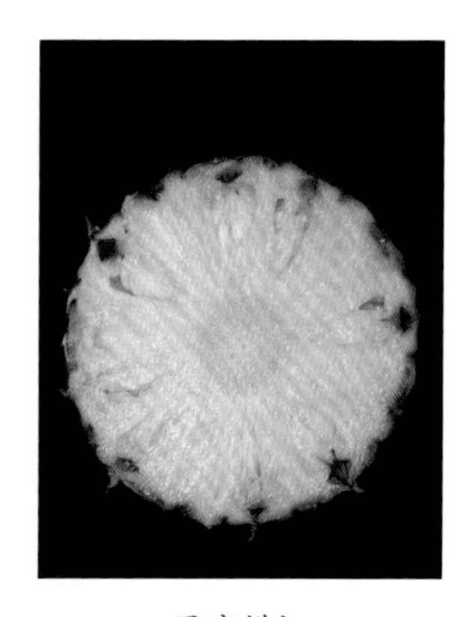

果实横切
Fruit transverse section

广西野种

编号：ACCESSION000126
种质名称：广西野种
原产地：不详
资源类型：国内收集
主要用途：育种
种质来源地：2021 年中国热带农业科学院南亚热带作物研究所从广西壮族自治区收集
植株姿态：开张
定植期：2021 年 2 月下旬
现红期：2023 年 3 月上旬
营养生长期：约 740 d
初花期：2023 年 3 月下旬
花开放时间：约 16 d
成熟期：2023 年 7 月下旬至 8 月上旬
果实发育期：约 125 d
果实形状：圆筒形
单果质量：832.9 g
纵径：12.3 cm
横径：10.5 cm
果形指数：1.2
未成熟果实果皮颜色：浅褐色
成熟果实果皮颜色：黄色，带绿斑
果颈：无
果基：平
果顶：平
果实小果能否剥离：不可剥离
果眼外观：扁平或微凹
果眼深度：深
果眼排列方式：右旋
果瘤：无
果肉颜色：淡黄色
果实香味：无
果实风味：酸至极酸
果肉质地：滑
果实外观综合评价：优
果实品质综合评价：差
果肉可溶性固形物含量：12.4%
果实成熟特性：晚熟
成熟期的一致性：不一致
丰产性：一般

Guangxi Wild Cultivar

Code: ACCESSION000126
Name: Guangxi Wild Cultivar
Place of origin: unknown
Resource type: domestically collected variety
Main ways of consumption: breeding
Material resource: SSCRI, CATAS collected from Guangxi Zhuang Autonomous Region in 2021
Plant posture: spreading
Planting time: late February in 2021
Open heart stage: early March in 2023
Vegetative growth period: approx. 740 d
Initial flowering time: late March in 2023
Flowering period: approx. 16 d
Fruit maturation time: late July to early August in 2023
Fruit developing period: approx. 125 d
Fruit shape: cylindrical
Single fruit weight: 832.9 g
Longitudinal diameter: 12.3 cm
Transverse diameter: 10.5 cm
Fruit shape index: 1.2
Immature fruit peel color: light brown
Mature fruit peel color: yellow with green stripe
Fruit neck: absent
Fruit base: flat
Fruit top: flat
Fruitlets adhesion situation: inseparable
Fruit eyes appearance: flat or slightly concave
Fruit eyes depth: deep
Fruit eyes arrangement: dextrorotation
Fruit tumors: absent
Flesh color: light yellow
Fruit aroma: no scent
Fruit flavour: sour or extreme sour
Flesh texture: smooth
Fruit appearance comprehensive evaluation: excellent
Fruit quality comprehensive evaluation: poor
Soluble solids content in flesh: 12.4%
Fruit ripening characteristics: medium mature
Maturity consistency: nonuniform
Productivity: normal level

植株
Plant

现红
Open heart

花
Flower

花序
Inflorescence

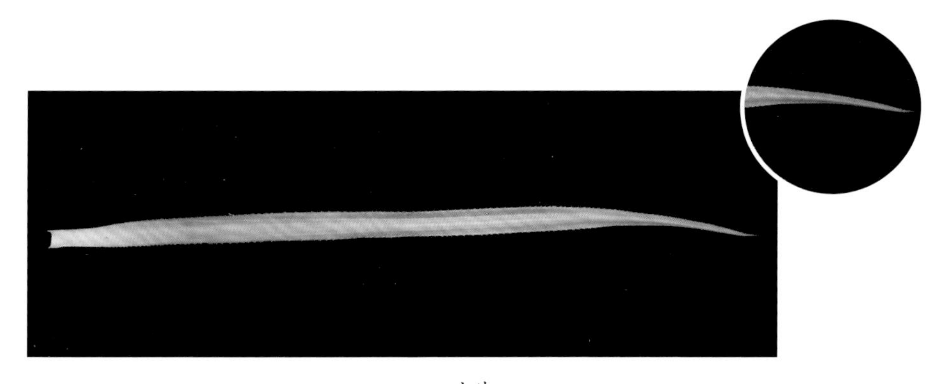

叶片
Leaf

带冠芽果
Fruit with crown bud

冠芽叶刺
Leaf spines in crown bud

果实
Fruit

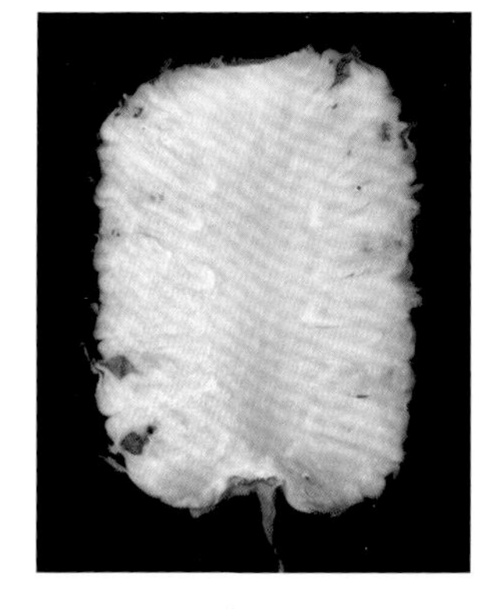

果实纵切
Fruit longitudinal section

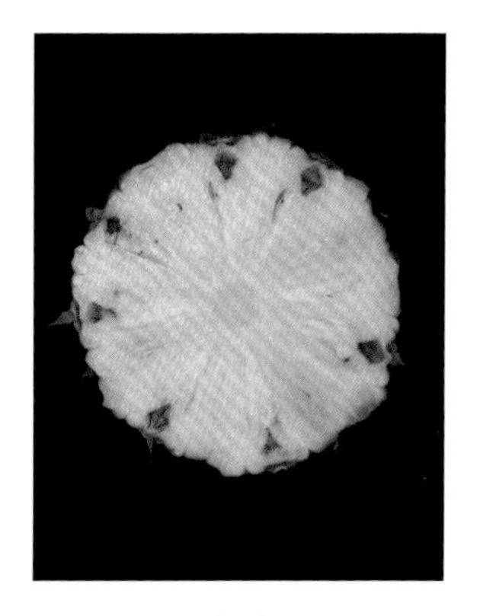

果实横切
Fruit transverse section

第三章
国外菠萝种质资源

Chapter 3
Pineapple Germplasm Resources from Abroad

Hilo

编号：ACCESSION000007
种质名称：Hilo
原产地：美国
资源类型：引进品种
主要用途：鲜食或加工
种质来源地：2008 年中国热带农业科学院南亚热带作物研究所从美国引进
植株姿态：开张
定植期：2022 年 1 月上旬
现红期：2023 年 3 月中旬
营养生长期：约 435 d
初花期：2023 年 4 月中下旬
花开放时间：约 21 d
成熟期：2023 年 7 月下旬
果实发育期：约 95 d
果实形状：圆筒形
单果质量：1192.6 g
纵径：14.3 cm
横径：11.4 cm
果形指数：1.3
未成熟果实果皮颜色：暗绿色
成熟果实果皮颜色：金黄 / 鲜黄色
果颈：无
果基：突起
果顶：平
果实小果能否剥离：可剥离
果眼外观：扁平或微凹
果眼深度：浅
果眼排列方式：左旋或右旋
果瘤：无
果肉颜色：淡黄色
果实香味：清香 / 微香
果实风味：微香甜
果肉质地：滑
果实外观综合评价：中
果实品质综合评价：中
果肉可溶性固形物含量：13.4%
果实成熟特性：中熟
成熟期的一致性：基本一致
丰产性：丰产

Hilo

Code: ACCESSION00007
Name: Hilo
Place of origin: America
Resource type: introduced variety
Main ways of consumption: fresh or processing
Initially introduction place: SSCRI, CATAS introduced from America in 2008
Plant posture: spreading
Planting time: early January in 2022
Open heart stage: mid-March in 2023
Vegetative growth period: approx. 435 d
Initial flowering time: mid to late April in 2023
Flowering period: approx. 21 d
Fruit maturation time: late July in 2023
Fruit developing period: approx. 95 d
Fruit shape: cylindrical
Single fruit weight: 1192.6 g
Longitudinal diameter: 14.3 cm
Transverse diameter: 11.4 cm
Fruit shape index: 1.3
Immature fruit peel color: dark green
Mature fruit peel color: golden yellow/vivid yellow
Fruit neck: absent
Fruit base: protrusive
Fruit top: flat
Fruitlets adhesion situation: separable
Fruit eyes appearance: flat or slightly concave
Fruit eyes depth: shallow
Fruit eyes arrangement: levorotation or dextrorotation
Fruit tumors: absent
Flesh color: pale yellow
Fruit aroma: faint/slight scent
Fruit flavour: slightly fragrant and sweet
Flesh texture: smooth
Fruit appearance comprehensive evaluation: medium
Fruit quality comprehensive evaluation: medium
Soluble solids content in flesh: 13.4%
Fruit ripening characteristics: medium mature
Maturity consistency: basically uniform
Productivity: productive

植株
Plant

现红
Open heart

花
Flower

花序
Inflorescence

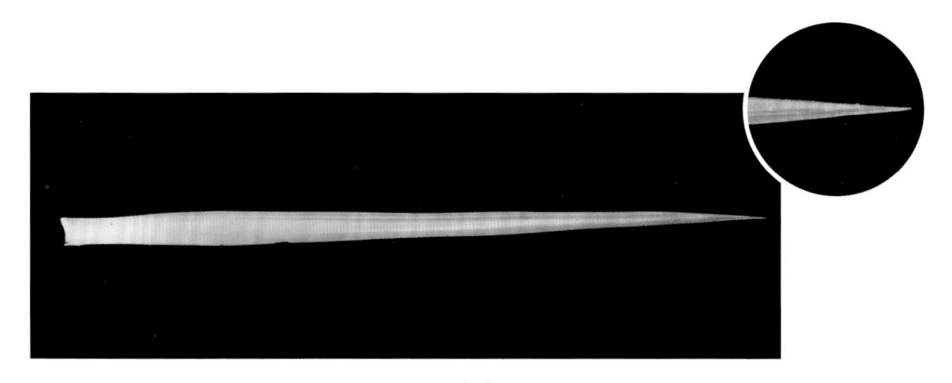

叶片
Leaf

带冠芽果
Fruit with crown bud

冠芽叶刺
Leaf spines in crown bud

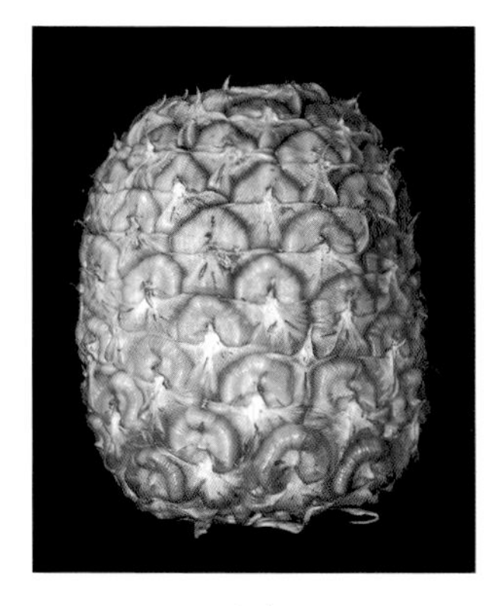

果实
Fruit

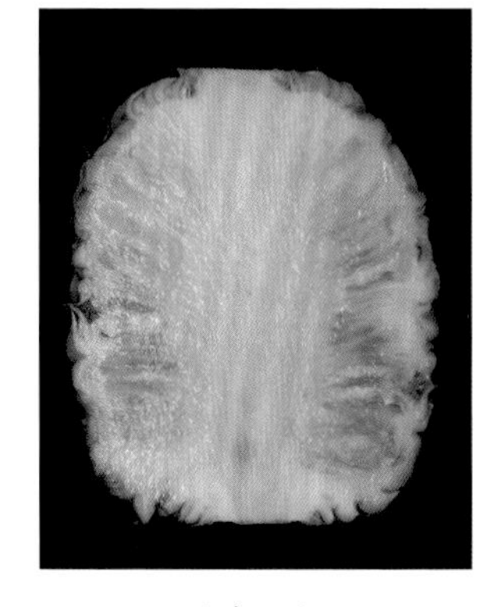

果实纵切
Fruit longitudinal section

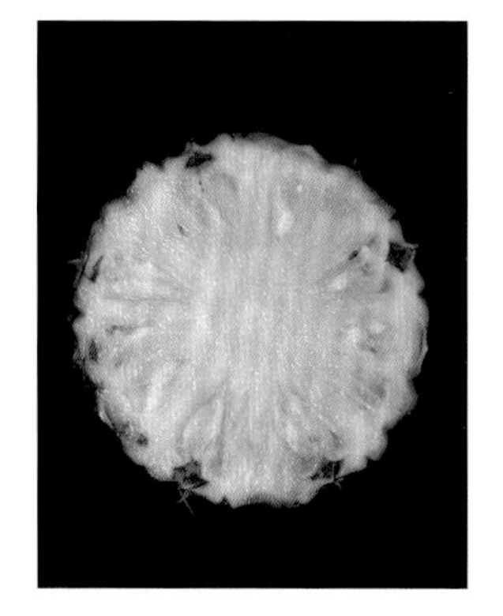

果实横切
Fruit transverse section

澳大利亚卡因

编号：ACCESSION000064
种质名称：澳大利亚卡因
原产地：澳大利亚
资源类型：引进品种
主要用途：鲜食或加工
种质来源地：2009年中国热带农业科学院南亚热带作物研究所从澳大利亚引进
植株姿态：开张
定植期：2021年2月下旬
现红期：2022年4月上旬
营养生长期：约410 d
初花期：2022年3月下旬至4月上旬
花开放时间：约20 d
成熟期：2022年7月下旬
果实发育期：约125 d
果实形状：圆筒形
单果质量：843.7 g
纵径：12.6 cm
横径：10.4 cm
果形指数：1.2
未成熟果实果皮颜色：淡绿/绿色
成熟果实果皮颜色：金黄/鲜黄色
果颈：无
果基：平
果顶：平
果实小果能否剥离：不可剥离
果眼外观：扁平或微凹
果眼深度：浅
果眼排列方式：左旋
果瘤：无或少
果肉颜色：淡黄色
果实香味：清香/微香
果实风味：甜酸
果肉质地：粗糙
果实外观综合评价：中
果实品质综合评价：中
果肉可溶性固形物含量：14.4%
果实成熟特性：晚熟
成熟期的一致性：基本一致
丰产性：一般

Australia Cayenne

Code: ACCESSION00064
Name: Australia Cayenne
Place of origin: Australia
Resource type: introduced variety
Main ways of consumption: fresh or processing
Initially introduction place: SSCRI, CATAS introduced from Australia in 2009
Plant posture: spreading
Planting time: late February in 2021
Open heart stage: early April in 2022
Vegetative growth period: approx. 410 d
Initial flowering time: late March to early April in 2022
Flowering period: approx. 20 d
Fruit maturation time: late July in 2022
Fruit developing period: approx. 125 d
Fruit shape: cylindrical
Single fruit weight: 843.7 g
Longitudinal diameter: 12.6 cm
Transverse diameter: 10.4 cm
Fruit shape index: 1.2
Immature fruit peel color: light green/green
Mature fruit peel color: golden yellow/vivid yellow
Fruit neck: absent
Fruit base: flat
Fruit top: flat
Fruitlets adhesion situation: inseparable
Fruit eyes appearance: flat or slightly concave
Fruit eyes depth: shallow
Fruit eyes arrangement: levorotation
Fruit tumors: absent or few
Flesh color: pale yellow
Fruit aroma: faint/slight scent
Fruit flavour: sweet-sour
Flesh texture: coarse
Fruit appearance comprehensive evaluation: medium
Fruit quality comprehensive evaluation: medium
Soluble solids content in flesh: 14.4%
Fruit ripening characteristics: late mature
Maturity consistency: basically uniform
Productivity: normal level

植株
Plant

现红
Open heart

花
Flower

花序
Inflorescence

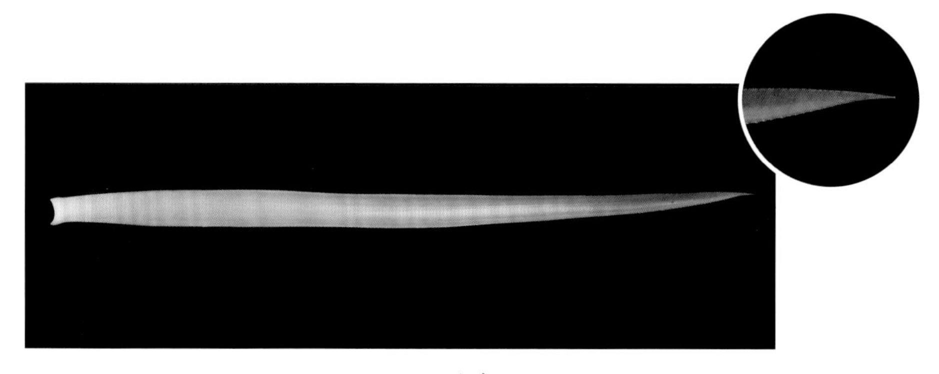

叶片
Leaf

带冠芽果
Fruit with crown bud

冠芽叶刺
Leaf spines in crown bud

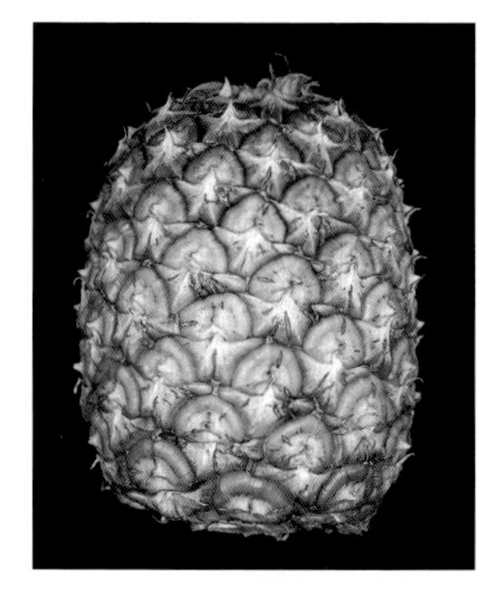

果实
Fruit

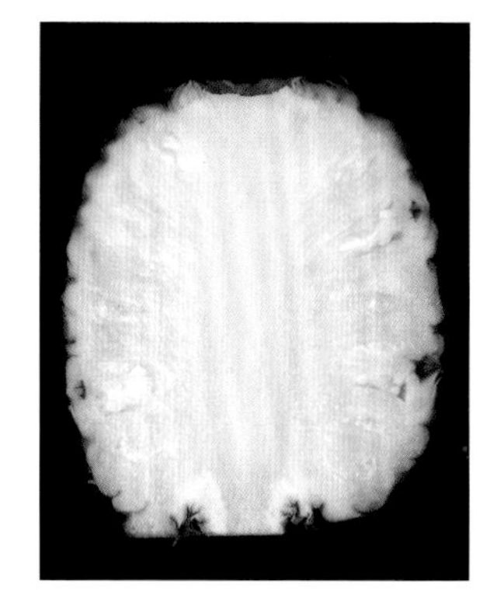

果实纵切
Fruit longitudinal section

果实横切
Fruit transverse section

Ripley

编号：ACCESSION000010
种质名称：Ripley
原产地：澳大利亚
资源类型：引进品种
主要用途：鲜食或加工
种质来源地：2008 年中国热带农业科学院南亚热带作物研究所从澳大利亚引进
植株姿态：开张
定植期：2021 年 2 月下旬
现红期：2022 年 2 月下旬
营养生长期：约 365 d
初花期：2022 年 3 月中旬
花开放时间：约 23 d
成熟期：2022 年 6 月中旬
果实发育期：约 90 d
果实形状：圆筒形
单果质量：584.9 g
纵径：13.3 cm
横径：8.8 cm
果形指数：1.5
未成熟果实果皮颜色：银绿色
成熟果实果皮颜色：金黄 / 鲜黄色
果颈：无
果基：平
果顶：平
果实小果能否剥离：可剥离
果眼外观：微隆起
果眼深度：较浅
果眼排列方式：右旋
果瘤：无
果肉颜色：金黄色
果实香味：芳香
果实风味：酸甜
果肉质地：粗糙
果实外观综合评价：中
果实品质综合评价：中
果肉可溶性固形物含量：18.1%
果实成熟特性：中熟
成熟期的一致性：一致
丰产性：低产

Ripley

Code: ACCESSION00010
Name: Ripley
Place of origin: Australia
Resource type: introduced variety
Main ways of consumption: fresh or processing
Initially introduction place: SSCRI, CATAS introduced from Australia in 2008
Plant posture: spreading
Planting time: late February in 2021
Open heart stage: late February in 2022
Vegetative growth period: approx. 365 d
Initial flowering time: mid-March in 2022
Flowering period: approx. 23 d
Fruit maturation time: mid-June in 2022
Fruit developing period: approx. 90 d
Fruit shape: cylindrical
Single fruit weight: 584.9 g
Longitudinal diameter: 13.3 cm
Transverse diameter: 8.8 cm
Fruit shape index: 1.5
Immature fruit peel color: silver green
Mature fruit peel color: golden yellow/ vivid yellow
Fruit neck: absent
Fruit base: flat
Fruit top: flat
Fruitlets adhesion situation: separable
Fruit eyes appearance: slightly embossed
Fruit eyes depth: relatively shallow
Fruit eyes arrangement: dextrorotation
Fruit tumors: absent
Flesh color: golden yellow
Fruit aroma: sweet fragrance
Fruit flavour: sour- sweet
Flesh texture: coarse
Fruit appearance comprehensive evaluation: medium
Fruit quality comprehensive evaluation: medium
Soluble solids content in flesh: 18.1%
Fruit ripening characteristics: medium mature
Maturity consistency: uniform
Productivity: low

植株
Plant

现红
Open heart

花
Flower

花序
Inflorescence

叶片
Leaf

带冠芽果
Fruit with crown bud

冠芽叶刺
Leaf spines in crown bud

果实
Fruit

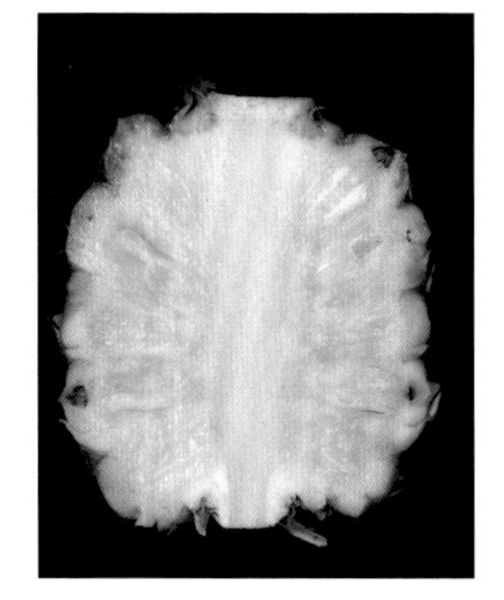

果实纵切
Fruit longitudinal section

果实横切
Fruit transverse section

澳大利亚引未知名

编号：ACCESSION000084
种质名称：澳大利亚引未知名
原产地：澳大利亚
资源类型：引进品种
主要用途：鲜食或加工
种质来源地：2010 年中国热带农业科学院南亚热带作物研究所从澳大利亚引进
植株姿态：开张
定植期：2022 年 1 月上旬
现红期：2023 年 2 月下旬至 3 月
营养生长期：约 425 d
初花期：2023 年 3 月下旬至 4 月上旬
花开放时间：约 19 d
成熟期：2023 年 7 月下旬
果实发育期：约 115 d
果实形状：圆筒形
单果质量：1353.7 g
纵径：14.8 cm
横径：11.9 cm
果形指数：1.2
未成熟果实果皮颜色：绿色
成熟果实果皮颜色：深黄至橙色
果颈：无
果基：平
果顶：浑圆
果实小果能否剥离：不可剥离
果眼外观：微隆起
果眼深度：浅
果眼排列方式：左旋
果瘤：无
果肉颜色：黄色
果实香味：无
果实风味：酸甜
果肉质地：滑
果实外观综合评价：优
果实品质综合评价：中
果肉可溶性固形物含量：13.4%
果实成熟特性：晚熟
成熟期的一致性：不一致
丰产性：丰产

Unknown Culitivar from Australia

Code: ACCESSION00084
Name: Unknown Culitivar from Australia
Place of origin: Australia
Resourcc typc: introduced variety
Main ways of consumption: fresh or processing
Initially introduction place: SSCRI, CATAS introduced from Australia in 2010
Plant posture: spreading
Planting time: early January in 2022
Open heart stage: late February to March in 2023
Vegetative growth period: approx. 425 d
Initial flowering time: late March to early April in 2023
Flowering period: approx. 19 d
Fruit maturation time: late July in 2023
Fruit developing period: approx. 115 d
Fruit shape: cylindrical
Single fruit weight: 1353.7 g
Longitudinal diameter: 14.8 cm
Transverse diameter: 11.9 cm
Fruit shape index: 1.2
Immature fruit peel color: green
Mature fruit peel color: deep yellow to orange
Fruit neck: absent
Fruit base: flat
Fruit top: perfectly round
Fruitlets adhesion situation: inseparable
Fruit eyes appearance: slightly embossed
Fruit eyes depth: shallow
Fruit eyes arrangement: levorotation
Fruit tumors: absent
Flesh color: yellow
Fruit aroma: no scent
Fruit flavour: sour-sweet
Flesh texture: smooth
Fruit appearance comprehensive evaluation: excellent
Fruit quality comprehensive evaluation: medium
Soluble solids content in flesh: 13.4%
Fruit ripening characteristics: late mature
Maturity consistency: nonuniform
Productivity: productive

植株
Plant

现红
Open heart

花
Flower

花序
Inflorescence

叶片
Leaf

带冠芽果
Fruit with crown bud

冠芽叶刺
Leaf spines in crown bud

果实
Fruit

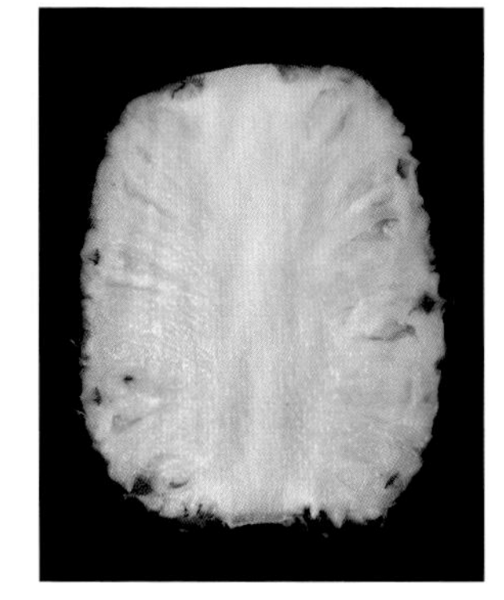

果实纵切
Fruit longitudinal section

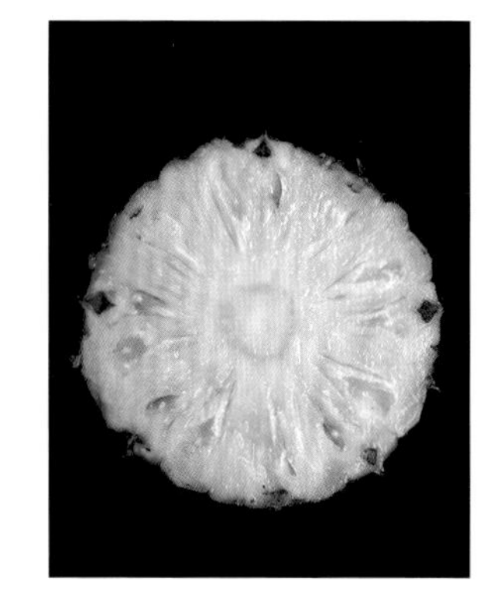

果实横切
Fruit transverse section

Sriracha

编号：ACCESSION000029
种质名称：Sriracha
原产地：泰国
资源类型：引进品种
主要用途：鲜食或加工
种质来源地：2008 年中国热带农业科学院南亚热带作物研究所从泰国引进
植株姿态：开张
定植期：2021 年 12 月下旬
现红期：2023 年 2 月下旬至 3 月上旬
营养生长期：约 430 d
初花期：2023 年 3 月下旬至 4 月上旬
花开放时间：约 26 d
成熟期：2023 年 7 月上旬
果实发育期：约 85 d
果实形状：圆筒形
单果质量：1024.0 g
纵径：13.3 cm
横径：10.5 cm
果形指数：1.3
未成熟果实果皮颜色：暗墨绿
成熟果实果皮颜色：深黄至橙色
果颈：无
果基：平
果顶：平
果实小果能否剥离：不可剥离
果眼外观：扁平或微凹
果眼深度：浅
果眼排列方式：右旋
果瘤：无
果肉颜色：淡黄色
果实香味：无
果实风味：清甜
果肉质地：滑
果实外观综合评价：中
果实品质综合评价：中
果肉可溶性固形物含量：13.9%
果实成熟特性：中熟
成熟期的一致性：一致
丰产性：一般

Sriracha

Code: ACCESSION000029
Name: Sriracha
Place of origin: Thailand
Resource type: introduced variety
Main ways of consumption: fresh or processing
Initially introduction place: SSCRI, CATAS introduced from Thailand in 2008
Plant posture: spreading
Planting time: late December in 2021
Open heart stage: late February to early March in 2023
Vegetative growth period: approx. 430 d
Initial flowering time: late March to early April in 2023
Flowering period: approx. 26 d
Fruit maturation time: early July in 2023
Fruit developing period: approx. 85 d
Fruit shape: cylindrical
Single fruit weight: 1024.0 g
Longitudinal diameter: 13.3 cm
Transverse diameter: 10.5 cm
Fruit shape index: 1.3
Immature fruit peel color: dark blackish green
Mature fruit peel color: deep yellow to orange
Fruit neck: absent
Fruit base: flat
Fruit top: flat
Fruitlets adhesion situation: inseparable
Fruit eyes appearance: flat or slightly concave
Fruit eyes depth: shallow
Fruit eyes arrangement: dextrorotation
Fruit tumors: absent
Flesh color: light yellow
Fruit aroma: no scent
Fruit flavour: mildly sweet
Flesh texture: smooth
Fruit appearance comprehensive evaluation: medium
Fruit quality comprehensive evaluation: medium
Soluble solids content in flesh: 13.9%
Fruit ripening characteristics: medium mature
Maturity consistency: uniform
Productivity: normal level

植株
Plant

现红
Open heart

花
Flower

花序
Inflorescence

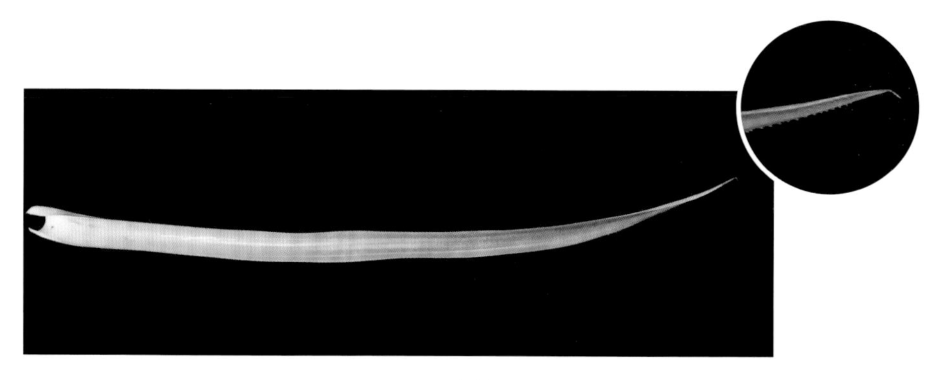

叶片
Leaf

带冠芽果
Fruit with crown bud

冠芽叶刺
Leaf spines in crown bud

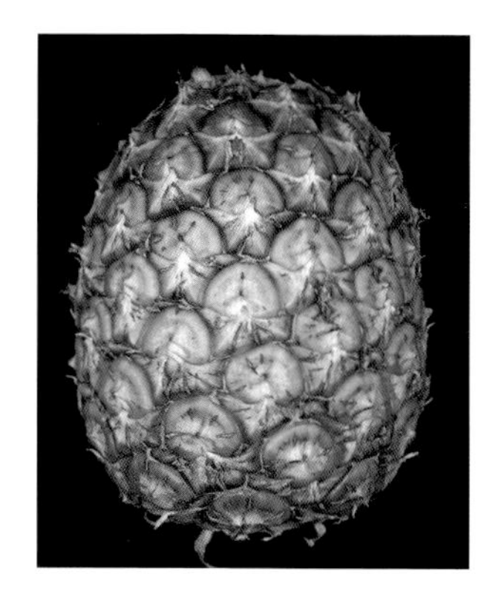

果实
Fruit

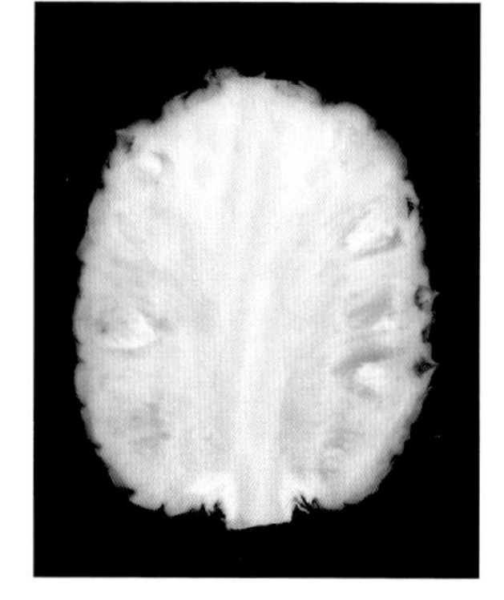

果实纵切
Fruit longitudinal section

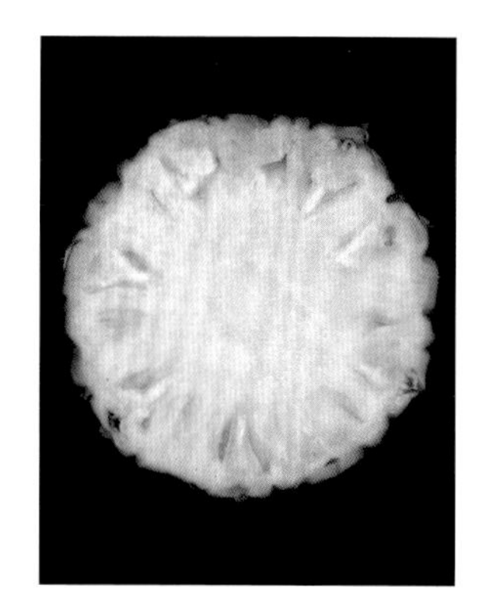

果实横切
Fruit transverse section

Nanglae

编号：ACCESSION000024
种质名称：Nanglae
原产地：泰国
资源类型：引进品种
主要用途：鲜食或加工
种质来源地：2008 年中国热带农业科学院南亚热带作物研究所从泰国引进
植株姿态：开张
定植期：2021 年 2 月下旬
现红期：2022 年 3 月上中旬
营养生长期：约 385 d
初花期：2022 年 4 月上旬
花开放时间：约 22 d
成熟期：2022 年 6 月上中旬
果实发育期：约 65 d
果实形状：短圆筒形
单果质量：673.2 g
纵径：11.6 cm
横径：10.0 cm
果形指数：1.2
未成熟果实果皮颜色：暗绿
成熟果实果皮颜色：暗黄 / 深黄色
果颈：无
果基：平
果顶：钝圆
果实小果能否剥离：不可剥离
果眼外观：微隆起
果眼深度：浅
果眼排列方式：右旋
果瘤：无
果肉颜色：淡黄色
果实香味：清香 / 微香
果实风味：甜酸
果肉质地：脆 / 爽脆
果实外观综合评价：差
果实品质综合评价：好
果肉可溶性固形物含量：19.9%
果实成熟特性：早熟
成熟期的一致性：基本一致
丰产性：低产

Nanglae

Code: ACCESSION000024
Name: Nanglae
Place of origin: Thailand
Resource type: introduced variety
Main ways of consumption: fresh or processing
Initially introduction place: SSCRI, CATAS introduced from Thailand in 2008
Plant posture: spreading
Planting time: late February in 2021
Open heart stage: early-to-mid March in 2022
Vegetative growth period: approx. 385 d
Initial flowering time: early April in 2022
Flowering period: approx. 22 d
Fruit maturation time: early-to-mid June in 2022
Fruit developing period: approx. 65 d
Fruit shape: short cylindrical
Single fruit weight: 673.2 g
Longitudinal diameter: 11.6 cm
Transverse diameter: 10.0 cm
Fruit shape index: 1.2
Immature fruit peel color: dark green
Mature fruit peel color: dark /deep yellow
Fruit neck: absent
Fruit base: flat
Fruit top: bluntly round
Fruitlets adhesion situation: inseparable
Fruit eyes appearance: slightly embossed
Fruit eyes depth: shallow
Fruit eyes arrangement: dextrorotation
Fruit tumors: absent
Flesh color: light yellow
Fruit aroma: faint/slight scent
Fruit flavour: sweet-sour
Flesh texture: crisp/crunchy
Fruit appearance comprehensive evaluation: poor
Fruit quality comprehensive evaluation: good
Soluble solids content in flesh: 19.9%
Fruit ripening characteristics: early mature
Maturity consistency: basically uniform
Productivity: low

植株
Plant

现红
Open heart

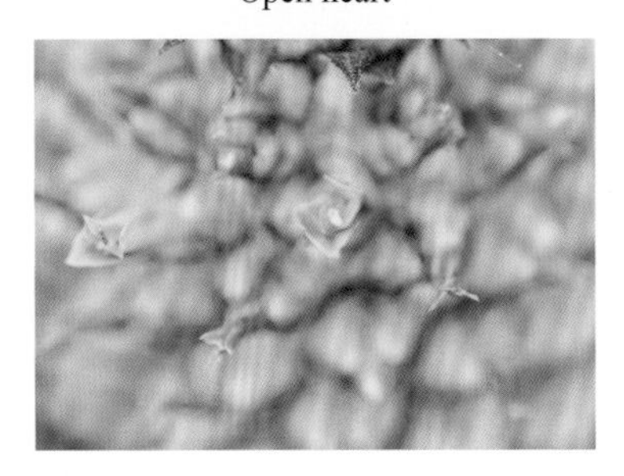

花
Flower

花序
Inflorescence

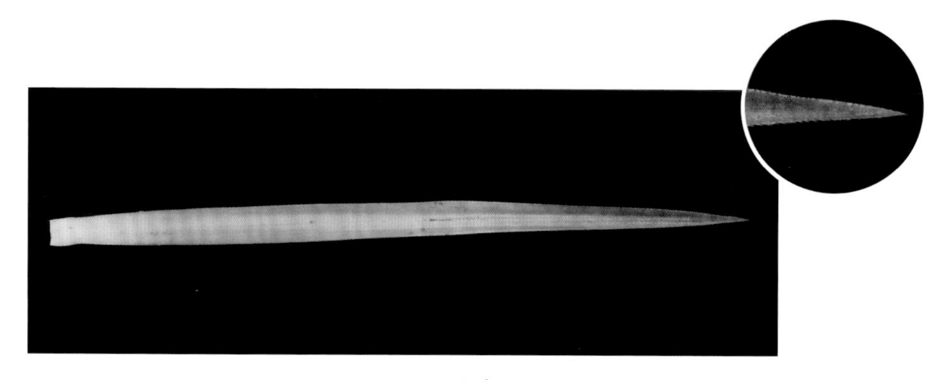

叶片
Leaf

带冠芽果
Fruit with crown bud

冠芽叶刺
Leaf spines in crown bud

果实
Fruit

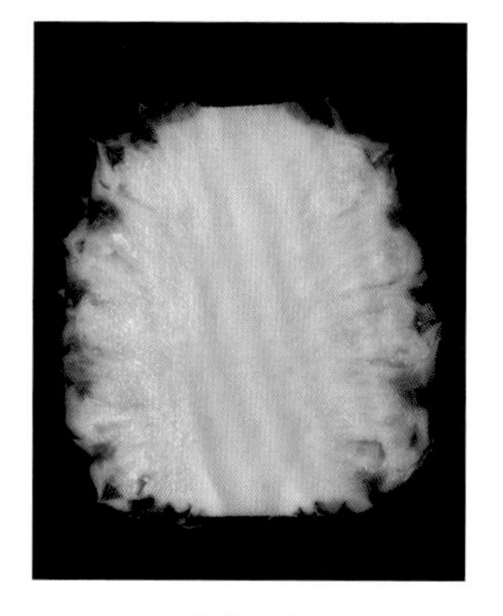

果实纵切
Fruit longitudinal section

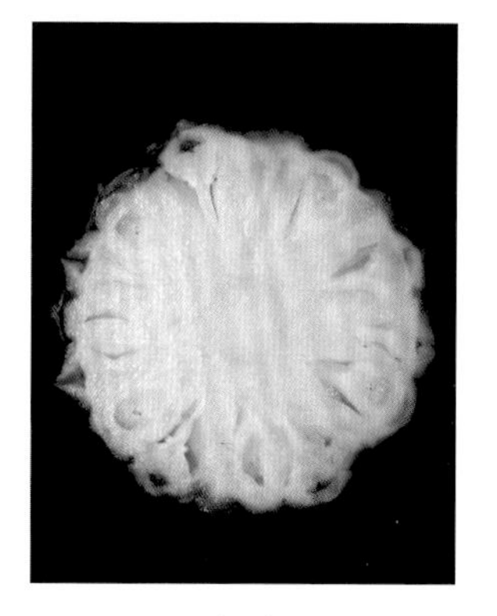

果实横切
Fruit transverse section

Pattavia

编号：ACCESSION000031
种质名称：Pattavia
原产地：泰国
资源类型：引进品种
主要用途：鲜食或加工
种质来源地：2006年中国热带农业科学院南亚热带作物研究所从泰国引进
植株姿态：开张
定植期：2021年2月下旬
现红期：2022年4月上旬
营养生长期：约410 d
初花期：2022年4月上旬
花开放时间：约26 d
成熟期：2022年8月中旬
果实发育期：约135 d
果实形状：圆筒形
单果质量：850.1 g
纵径：12.6 cm
横径：11.7 cm
果形指数：1.1
未成熟果实果皮颜色：淡绿/绿色
成熟果实果皮颜色：金黄/鲜黄色
果颈：无
果基：平
果顶：平
果实小果能否剥离：不可剥离
果眼外观：扁平或微凹
果眼深度：浅
果眼排列方式：左旋
果瘤：无
果肉颜色：黄色
果实香味：无
果实风味：清甜
果肉质地：粗糙
果实外观综合评价：好
果实品质综合评价：中
果肉可溶性固形物含量：14.3%
果实成熟特性：晚熟
成熟期的一致性：不一致
丰产性：一般

Pattavia

Code: ACCESSION000031
Name: Pattavia
Place of origin: Thailand
Resource type: introduced variety
Main ways of consumption: fresh or processing
Initially introduction place: SSCRI, CATAS introduced from Thailand in 2006
Plant posture: spreading
Planting time: late February in 2021
Open heart stage: early April in 2022
Vegetative growth period: approx. 410 d
Initial flowering time: early April in 2022
Flowering period: approx. 26 d
Fruit maturation time: mid-August in 2022
Fruit developing period: approx. 135 d
Fruit shape: cylindrical
Single fruit weight: 850.1 g
Longitudinal diameter: 12.6 cm
Transverse diameter: 11.7 cm
Fruit shape index: 1.1
Immature fruit peel color: light green/green
Mature fruit peel color: golden yellow/vivid yellow
Fruit neck: absent
Fruit base: flat
Fruit top: flat
Fruitlets adhesion situation: inseparable
Fruit eyes appearance: flat or slightly concave
Fruit eyes depth: shallow
Fruit eyes arrangement: levorotation
Fruit tumors: few
Flesh color: yellow
Fruit aroma: no scent
Fruit flavour: mildly sweet
Flesh texture: coarse
Fruit appearance comprehensive evaluation: good
Fruit quality comprehensive evaluation: medium
Soluble solids content in flesh: 14.3%
Fruit ripening characteristics: late mature
Maturity consistency: nonuniform
Productivity: normal level

植株
Plant

现红
Open heart

花
Flower

花序
Inflorescence

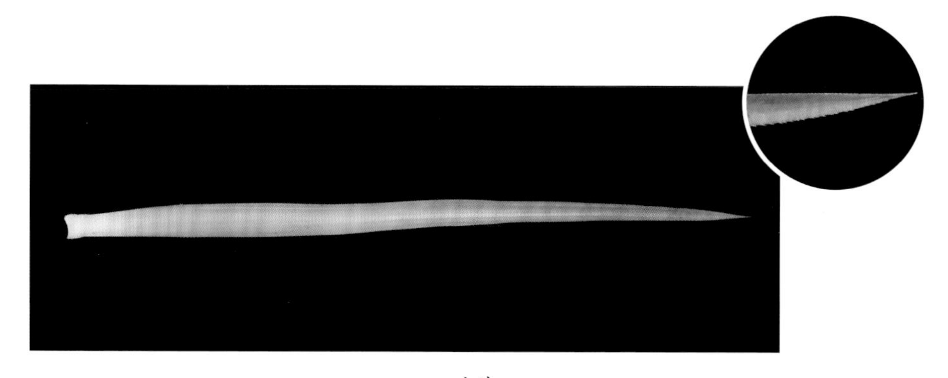

叶片
Leaf

带冠芽果
Fruit with crown bud

冠芽叶刺
Leaf spines in crown bud

果实
Fruit

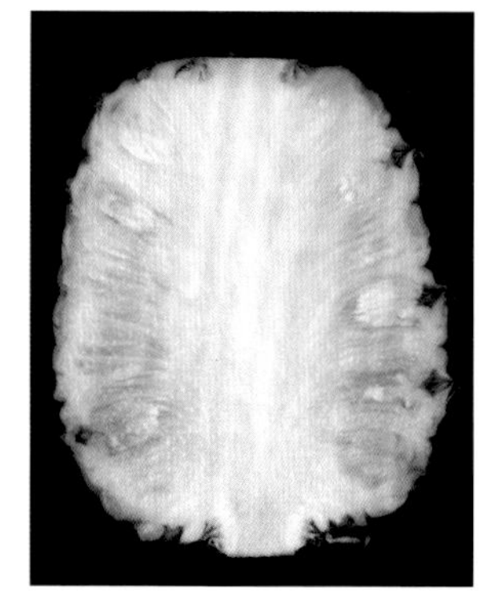

果实纵切
Fruit longitudinal section

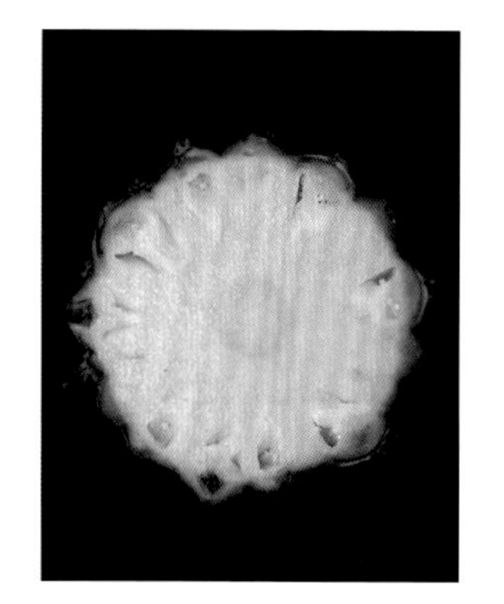

果实横切
Fruit transverse section

Pulae

编号：ACCESSION000073
种质名称：Pulae
原产地：泰国
资源类型：引进品种
主要用途：鲜食或加工
种质来源地：2009 年中国热带农业科学院南亚热带作物研究所从泰国引进
植株姿态：开张
定植期：2022 年 1 月上旬
现红期：2023 年 2 月下旬至 3 月上旬
营养生长期：约 430 d
初花期：2023 年 3 月下旬至 4 月上旬
花开放时间：约 16 d
成熟期：2023 年 6 月中旬
果实发育期：约 75 d
果实形状：短圆筒形
单果质量：595.7 g
纵径：11.1 cm
横径：9.7 cm
果形指数：1.1
未成熟果实果皮颜色：暗绿
成熟果实果皮颜色：亮黄
果颈：无
果基：平
果顶：平
果实小果能否剥离：可剥离
果眼外观：突起 / 隆起
果眼深度：较浅
果眼排列方式：右旋
果瘤：无
果肉颜色：黄色
果实香味：清香 / 微香
果实风味：甜酸
果肉质地：脆 / 爽脆
果实外观综合评价：中
果实品质综合评价：好
果肉可溶性固形物含量：19.0%
果实成熟特性：早熟
成熟期的一致性：一致
丰产性：低产

Pulae

Code: ACCESSION000073
Name: Pulae
Place of origin: Thailand
Resource type: introduced variety
Main ways of consumption: fresh or processing
Initially introduction place: SSCRI, CATAS introduced from Thailand in 2009
Plant posture: spreading
Planting time: early January in 2022
Open heart stage: late February to early March in 2023
Vegetative growth period: approx. 430 d
Initial flowering time: late March to early April in 2023
Flowering period: approx. 16 d
Fruit maturation time: mid-June in 2023
Fruit developing period: approx. 75 d
Fruit shape: short cylindrical
Single fruit weight: 595.7 g
Longitudinal diameter: 11.1 cm
Transverse diameter: 9.7 cm
Fruit shape index: 1.1
Immature fruit peel color: dark green
Mature fruit peel color: bright yellow
Fruit neck: absent
Fruit base: flat
Fruit top: flat
Fruitlets adhesion situation: separable
Fruit eyes appearance: convex/embossed
Fruit eyes depth: relatively shallow
Fruit eyes arrangement: dextrorotation
Fruit tumors: absent
Flesh color: yellow
Fruit aroma: faint/slight scent
Fruit flavour: sweet-sour
Flesh texture: crisp/crunchy
Fruit appearance comprehensive evaluation: medium
Fruit quality comprehensive evaluation: good
Soluble solids content in flesh: 19.0%
Fruit ripening characteristics: early mature
Maturity consistency: uniform
Productivity: low

植株
Plant

现红
Open heart

花
Flower

花序
Inflorescence

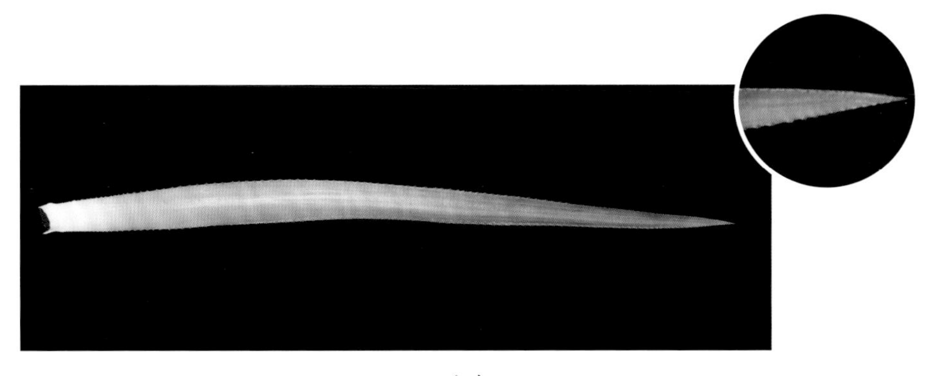

叶片
Leaf

带冠芽果
Fruit with crown bud

冠芽叶刺
Leaf spines in crown bud

果实
Fruit

果实纵切
Fruit longitudinal section

果实横切
Fruit transverse section

Pérola

编号：ACCESSION000033
种质名称：Pérola
原产地：巴西
资源类型：引进品种
主要用途：鲜食或加工
种质来源地：2008年中国热带农业科学院南亚热带作物研究所从巴西引进
植株姿态：开张
定植期：2021年2月下旬
现红期：2022年3月中旬
营养生长期：约390 d
初花期：2022年4月中旬
花开放时间：约21 d
成熟期：2022年6月下旬
果实发育期：约70 d
果实形状：长圆筒形
单果质量：694.2 g
纵径：12.1 cm
横径：9.0 cm
果形指数：1.3
未成熟果实果皮颜色：暗绿
成熟果实果皮颜色：黄色，带绿斑
果颈：有
果基：平
果顶：平
果实小果能否剥离：不可剥离
果眼外观：微隆起
果眼深度：浅
果眼排列方式：右旋
果瘤：无
果肉颜色：淡黄色
果实香味：无
果实风味：酸甜
果肉质地：粗糙
果实外观综合评价：中
果实品质综合评价：中
果肉可溶性固形物含量：14.4%
果实成熟特性：早熟
成熟期的一致性：一致
丰产性：低产

Pérola

Code: ACCESSION000033
Name: Pérola
Place of origin: Brazil
Resource type: introduced variety
Main ways of consumption: fresh or processing
Initially introduction place: SSCRI, CATAS introduced from Brazil in 2008
Plant posture: spreading
Planting time: late February in 2021
Open heart stage: mid-March in 2022
Vegetative growth period: approx. 390 d
Initial flowering time: mid-April in 2022
Flowering period: approx. 21 d
Fruit maturation time: late June in 2022
Fruit developing period: approx. 70 d
Fruit shape: long cylindrical
Single fruit weight: 694.2 g
Longitudinal diameter: 12.1 cm
Transverse diameter: 9.0 cm
Fruit shape index: 1.3
Immature fruit peel color: dark green
Mature fruit peel color: yellow with green stripe
Fruit neck: present
Fruit base: flat
Fruit top: flat
Fruitlets adhesion situation: inseparable
Fruit eyes appearance: slightly embossed
Fruit eyes depth: shallow
Fruit eyes arrangement: dextrorotation
Fruit tumors: absent
Flesh color: pale yellow
Fruit aroma: no scent
Fruit flavour: sour- sweet
Flesh texture: coarse
Fruit appearance comprehensive evaluation: medium
Fruit quality comprehensive evaluation: medium
Soluble solids content in flesh: 14.4%
Fruit ripening characteristics: early mature
Maturity consistency: uniform
Productivity: low

植株
Plant

现红
Open heart

花
Flower

花序
Inflorescence

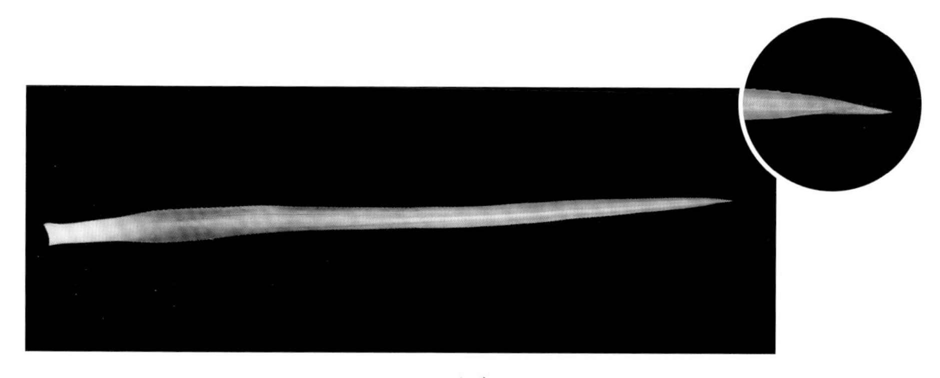

叶片
Leaf

带冠芽果
Fruit with crown bud

冠芽叶刺
Leaf spines in crown bud

果实
Fruit

果实纵切
Fruit longitudinal section

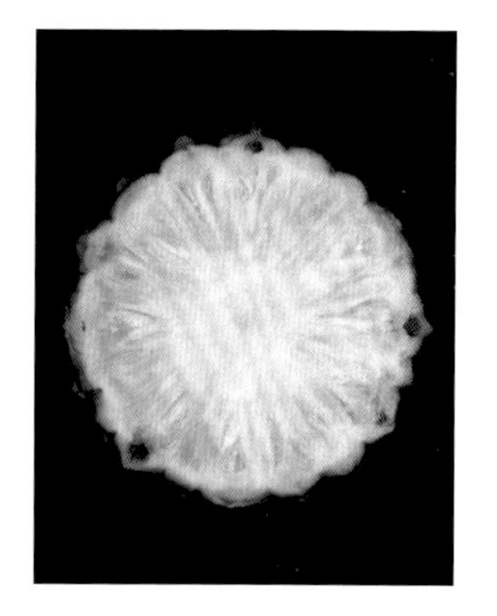

果实横切
Fruit transverse section

红色野生种

编号：ACCESSION000034
种质名称：红色野生种
原产地：巴西
资源类型：引进品种
主要用途：育种或观赏
种质来源地：2008 年中国热带农业科学院南亚热带作物研究所从巴西引进
植株姿态：开张
定植期：2021 年 12 月下旬
现红期：2023 年 2 月下旬
营养生长期：约 425 d
初花期：2023 年 4 月上旬
花开放时间：约 20 d
成熟期：2023 年 6 月中旬
果实发育期：约 70 d
果实形状：长圆筒形
单果质量：71.6 g
纵径：6.9 cm
横径：4.4 cm
果形指数：1.6
未成熟果实果皮颜色：暗紫红
成熟果实果皮颜色：浅紫红
果颈：无
果基：平
果顶：平
果实小果能否剥离：可剥离
果眼外观：扁平或微凹
果眼深度：浅
果眼排列方式：右旋
果瘤：无或少
果肉颜色：白色
果实香味：无
果实风味：微酸
果肉质地：粗糙
果实外观综合评价：差
果实品质综合评价：差
果肉可溶性固形物含量：8.6%
果实成熟特性：早熟
成熟期的一致性：一致
丰产性：低产

Red Wild Cultivar

Code: ACCESSION000034
Name: Red Wild Cultivar
Place of origin: Brazil
Resource type: introduced variety
Main ways of consumption: breeding or ornamentation
Initially introduction place: SSCRI, CATAS introduced from Thailand in 2008
Plant posture: spreading
Planting time: late December in 2021
Open heart stage: late February in 2023
Vegetative growth period: approx. 425 d
Initial flowering time: early Aprid in 2023
Flowering period: approx. 20 d
Fruit maturation time: mid-June in 2023
Fruit developing period: approx. 70 d
Fruit shape: long cylindrical
Single fruit weight: 71.6 g
Longitudinal diameter: 6.9 cm
Transverse diameter: 4.4 cm
Fruit shape index: 1.6
Immature fruit peel color: violet red
Mature fruit peel color: light violet red
Fruit neck: absent
Fruit base: flat
Fruit top: flat
Fruitlets adhesion situation: separable
Fruit eyes appearance: flat or slightly concave
Fruit eyes depth: shallow
Fruit eyes arrangement: dextrorotation
Fruit tumors: absent or few
Flesh color: white
Fruit aroma: no scent
Fruit flavour: slightly sour
Flesh texture: coarse
Fruit appearance comprehensive evaluation: poor
Fruit quality comprehensive evaluation: poor
Soluble solids content in flesh: 8.6%
Fruit ripening characteristics: early mature
Maturity consistency: uniform
Productivity: low

植株
Plant

现红
Open heart

花
Flower

花序
Inflorescence

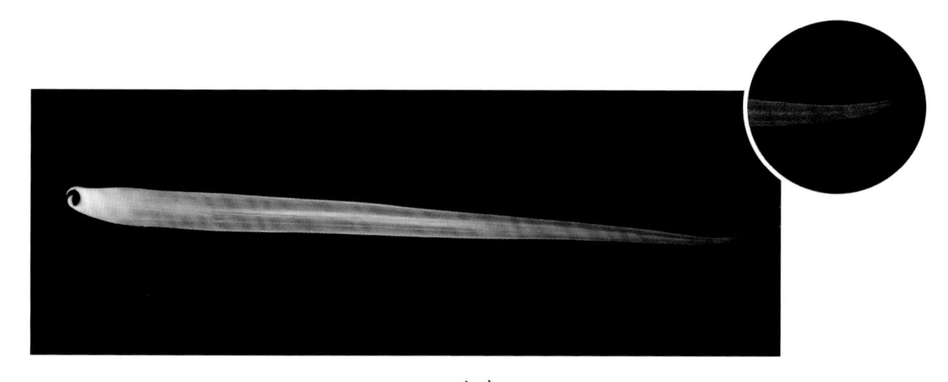

叶片
Leaf

带冠芽果
Fruit with crown bud

冠芽叶刺
Leaf spines in crown bud

果实
Fruit

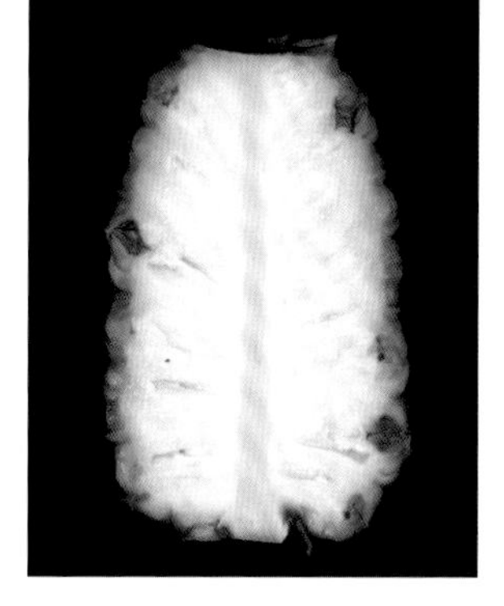

果实纵切
Fruit longitudinal section

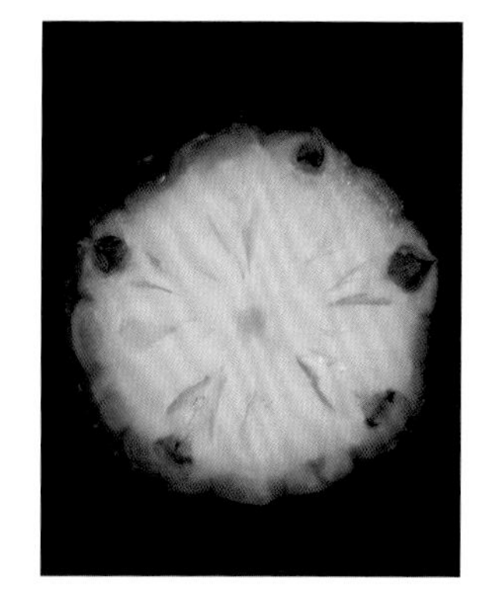

果实横切
Fruit transverse section

亚马逊野生种杂交后代 1

编号：ACCESSION000035
种质名称：亚马逊野生种杂交后代 1
原产地：巴西
资源类型：引进品种
主要用途：鲜食或加工
种质来源地：2008 年中国热带农业科学院南亚热带作物研究所从巴西引进
植株姿态：开张
定植期：2022 年 1 月上旬
现红期：2023 年 3 月上旬
营养生长期：约 430 d
初花期：2023 年 3 月下旬
花开放时间：约 10 d
成熟期：2023 年 7 月下旬至 8 月上旬
果实发育期：约 125 d
果实形状：长圆筒形
单果质量：578.1 g
纵径：12.7 cm
横径：8.7 cm
果形指数：1.5
未成熟果实果皮颜色：暗绿
成熟果实果皮颜色：金黄 / 鲜黄色
果颈：无
果基：弧形
果顶：钝圆
果实小果能否剥离：不可剥离
果眼外观：突起 / 隆起
果眼深度：浅
果眼排列方式：左旋
果瘤：无
果肉颜色：白色
果实香味：无
果实风味：酸甜
果肉质地：滑
果实外观综合评价：差
果实品质综合评价：中
果肉可溶性固形物含量：12.8%
果实成熟特性：晚熟
成熟期的一致性：基本一致
丰产性：低产

Hybrid No. 1 from Amazon Wild Species

Code: ACCESSION000035
Name: Hybrid No. 1 from Amazon Wild Species
Place of origin: Brazil
Resource type: introduced variety
Main ways of consumption: fresh or processing
Initially introduction place: SSCRI, CATAS introduced from Brazil in 2008
Plant posture: spreading
Planting time: early January in 2022
Open heart stage: early March in 2023
Vegetative growth period: approx. 430 d
Initial flowering time: late March in 2023
Flowering period: approx. 10 d
Fruit maturation time: late July to early August in 2023
Fruit developing period: approx. 125 d
Fruit shape: long cylindrical
Single fruit weight: 578.1 g
Longitudinal diameter: 12.7 cm
Transverse diameter: 8.7 cm
Fruit shape index: 1.5
Immature fruit peel color: dark green
Mature fruit peel color: golden yellow/vivid yellow
Fruit neck: absent
Fruit base: curved
Fruit top: bluntly round
Fruitlets adhesion situation: inseparable
Fruit eyes appearance: convex/embossed
Fruit eyes depth: shallow
Fruit eyes arrangement: levorotation
Fruit tumors: absent
Flesh color: white
Fruit aroma: no scent
Fruit flavour: sour-sweet
Flesh texture: smooth
Fruit appearance comprehensive evaluation: poor
Fruit quality comprehensive evaluation: medium
Soluble solids content in flesh: 12.8%
Fruit ripening characteristics: late mature
Maturity consistency: basically uniform
Productivity: low

植株
Plant

现红
Open heart

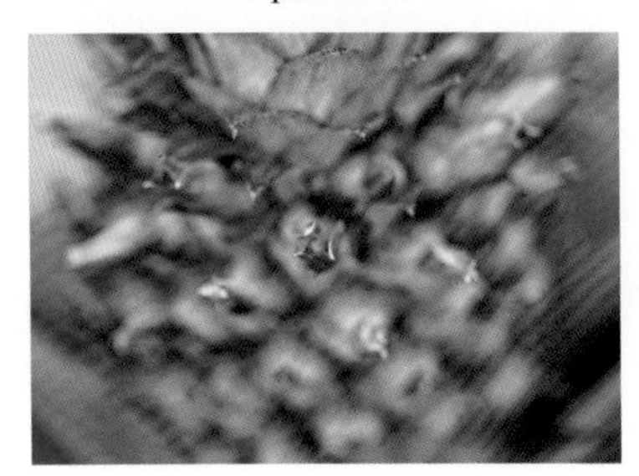

花
Flower

花序
Inflorescence

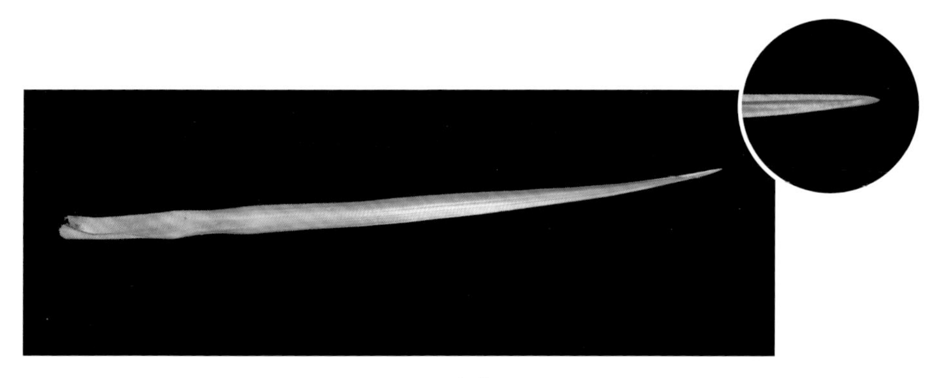

叶片
Leaf

带冠芽果
Fruit with crown bud

冠芽叶刺
Leaf spines in crown bud

果实
Fruit

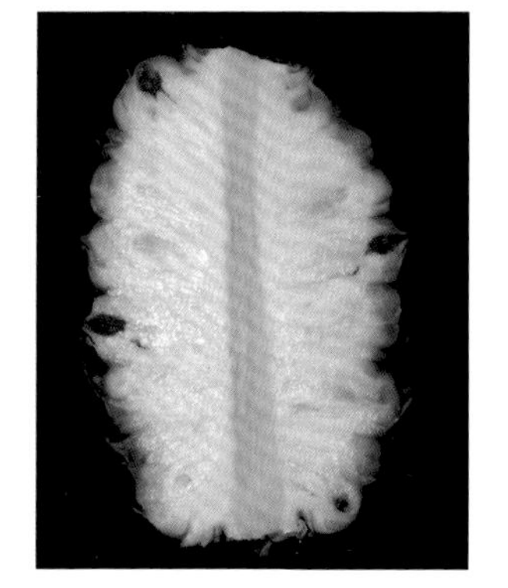

果实纵切
Fruit longitudinal section

果实横切
Fruit transverse section

亚马逊野生种杂交后代 2

编号：ACCESSION000036
种质名称：亚马逊野生种杂交后代 2
原产地：巴西
资源类型：引进品种
主要用途：鲜食或加工
种质来源地：2008 年中国热带农业科学院南亚热带作物研究所从巴西引进
植株姿态：开张
定植期：2021 年 12 月下旬
现红期：2023 年 2 月下旬
营养生长期：约 425 d
初花期：2023 年 3 月下旬至 4 月上旬
花开放时间：约 19 d
成熟期：2023 年 6 月中下旬
果实发育期：约 80 d
果实形状：长圆锥形
单果质量：1401.1 g
纵径：17.7 cm
横径：11.7 cm
果形指数：1.5
未成熟果实果皮颜色：暗绿
成熟果实果皮颜色：金黄 / 鲜黄色
果颈：无
果基：平
果顶：钝圆
果实小果能否剥离：不可剥离
果眼外观：突起 / 隆起
果眼深度：较浅
果眼排列方式：右旋
果瘤：无
果肉颜色：奶油色
果实香味：无
果实风味：酸甜
果肉质地：滑
果实外观综合评价：优
果实品质综合评价：中
果肉可溶性固形物含量：13.4%
果实成熟特性：中熟
成熟期的一致性：基本一致
丰产性：丰产

Hybrid No. 2 from Amazon Wild Species

Code: ACCESSION000036
Name: Hybrid No. 2 from Amazon Wild Species
Place of origin: Brazil
Resource type: introduced variety
Main ways of consumption: fresh or processing
Initially introduction place: SSCRI, CATAS introduced from Brazil in 2008
Plant posture: spreading
Planting time: late December in 2021
Open heart stage: late February in 2023
Vegetative growth period: approx. 425 d
Initial flowering time: late March to early April in 2023
Flowering period: approx. 19 d
Fruit maturation time: mid-to-late June in 2023
Fruit developing period: approx. 80 d
Fruit shape: long conical
Single fruit weight: 1401.1 g
Longitudinal diameter: 17.7 cm
Transverse diameter: 11.7 cm
Fruit shape index: 1.5
Immature fruit peel color: dark green
Mature fruit peel color: golden yellow/vivid yellow
Fruit neck: absent
Fruit base: flat
Fruit top: bluntly round
Fruitlets adhesion situation: inseparable
Fruit eyes appearance: convex/embossed
Fruit eyes depth: relatively shallow
Fruit eyes arrangement: dextrorotation
Fruit tumors: absent
Flesh color: cream
Fruit aroma: no
Fruit flavour: sour-sweet
Flesh texture: smooth
Fruit appearance comprehensive evaluation: excellent
Fruit quality comprehensive evaluation: medium
Soluble solids content in flesh: 13.4%
Fruit ripening characteristics: medium mature
Maturity consistency: basically uniform
Productivity: productive

植株
Plant

现红
Open heart

花
Flower

花序
Inflorescence

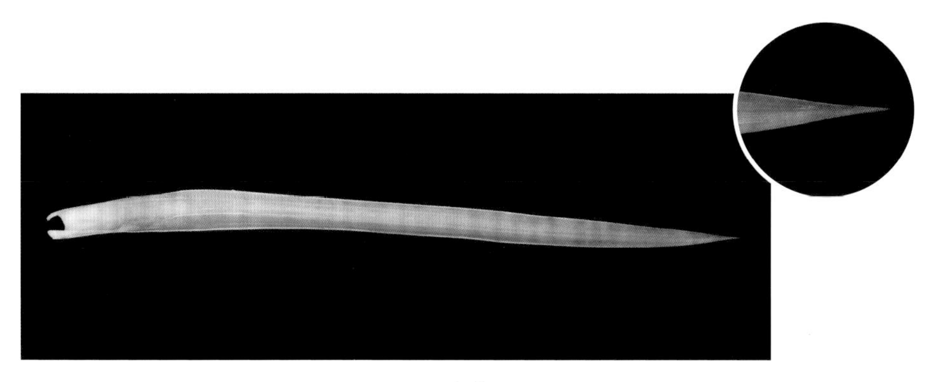

叶片
Leaf

带冠芽果
Fruit with crown bud

冠芽叶刺
Leaf spines in crown bud

果实
Fruit

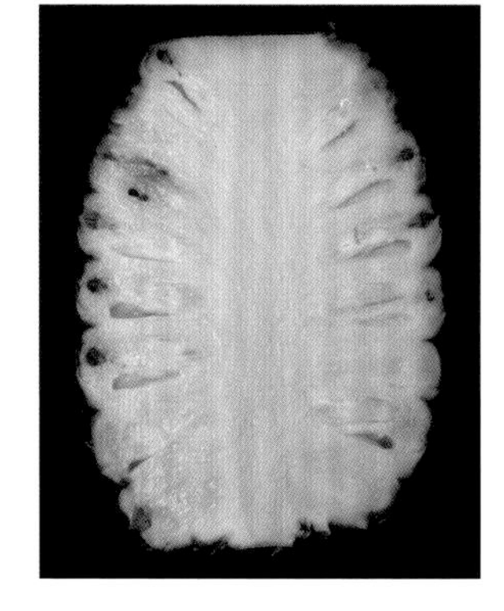

果实纵切
Fruit longitudinal section

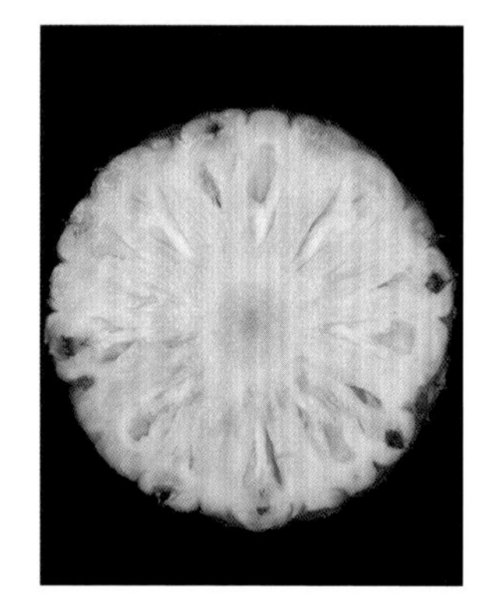

果实横切
Fruit transverse section

DL2

编号：ACCESSION000097
种质名称：DL2
原产地：越南
资源类型：引进品种
主要用途：鲜食或加工
种质来源地：2009年中国热带农业科学院南亚热带作物研究所从越南引进
植株姿态：开张
定植期：2022年1月上旬
现红期：2023年2月下旬
营养生长期：约425 d
初花期：2023年3月下旬
花开放时间：约16 d
成熟期：2023年6月中旬
果实发育期：约80 d
果实形状：圆筒形
单果质量：758.2 g
纵径：13.1 cm
横径：10.0 cm
果形指数：1.3
未成熟果实果皮颜色：银绿色
成熟果实果皮颜色：黄色，带绿斑
果颈：有
果基：弧形
果顶：平
果实小果能否剥离：不可剥离
果眼外观：突起 / 隆起
果眼深度：浅
果眼排列方式：右旋
果瘤：无或少
果肉颜色：黄色
果实香味：清香 / 微香
果实风味：甜酸
果肉质地：滑
果实外观综合评价：中
果实品质综合评价：好
果肉可溶性固形物含量：15.9%
果实成熟特性：晚熟
成熟期的一致性：基本一致
丰产性：低产

DL2

Code: ACCESSION000097
Name: DL2
Place of origin: Vietnam
Resource type: introduced variety
Main ways of consumption: fresh or processing
Initially introduction place: SSCRI, CATAS introduced from Vietnam in 2009.
Plant posture: spreading
Planting time: early January in 2022
Open heart stage: late February in 2023
Vegetative growth period: approx.425 d
Initial flowering time: late March in 2023
Flowering period: approx. 16 d
Fruit maturation time: mid-June in 2023
Fruit developing period: approx. 80 d
Fruit shape: cylindrical
Single fruit weight: 758.2 g
Longitudinal diameter: 13.1 cm
Transverse diameter: 10.0 cm
Fruit shape index: 1.3
Immature fruit peel color: silver green
Mature fruit peel color: yellow with green stripe
Fruit neck: present
Fruit base: curved
Fruit top: flat
Fruitlets adhesion situation: inseparable
Fruit eyes appearance: convex/embossed
Fruit eyes depth: shallow
Fruit eyes arrangement: dextrorotation
Fruit tumors: absent or few
Flesh color: yellow
Fruit aroma: faint/slight scent
Fruit flavour: sweet-sour
Flesh texture: smooth
Fruit appearance comprehensive evaluation: medium
Fruit quality comprehensive evaluation: good
Soluble solids content in flesh: 15.9%
Fruit ripening characteristics: late mature
Maturity consistency: basically uniform
Productivity: low

植株
Plant

现红
Open heart

花
Flower

花序
Inflorescence

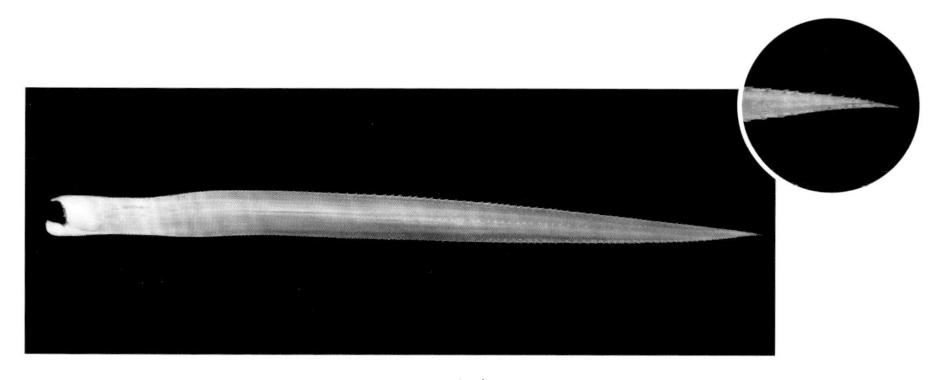

叶片
Leaf

带冠芽果
Fruit with crown bud

冠芽叶刺
Leaf spines in crown bud

果实
Fruit

果实纵切
Fruit longitudinal section

果实横切
Fruit transverse section

DN3

编号：ACCESSION000094
种质名称：DN3
原产地：越南
资源类型：引进品种
主要用途：鲜食或加工
种质来源地：2009 年中国热带农业科学院南亚热带作物研究所从越南引进
植株姿态：开张
定植期：2021 年 2 月中旬
现红期：2022 年 3 月中旬
营养生长期：约 390 d
初花期：2022 年 4 月中旬
花开放时间：约 27 d
成熟期：2022 年 7 月上中旬
果实发育期：约 95 d
果实形状：短圆筒形
单果质量：581.6 g
纵径：9.5 cm
横径：9.6 cm
果形指数：1.0
未成熟果实果皮颜色：暗墨绿色
成熟果实果皮颜色：浅褐 / 黄红色
果颈：无
果基：平
果顶：平
果实小果能否剥离：可剥离
果眼外观：突起 / 隆起
果眼深度：浅
果眼排列方式：左旋或右旋
果瘤：无或多
果肉颜色：淡黄色
果实香味：清香 / 微香
果实风味：浓甜
果肉质地：脆 / 爽脆
果实外观综合评价：差
果实品质综合评价：优
果肉可溶性固形物含量：19.7%
果实成熟特性：晚熟
成熟期的一致性：不一致
丰产性：低产

DN3

Code: ACCESSION000094
Name: DN3
Place of origin: Vietnam
Resource type: introduced variety
Main ways of consumption: fresh or processing
Initially introduction place: SSCRI, CATAS introduced from Vietnam in 2009
Plant posture: spreading
Planting time: mid-February in 2021
Open heart stage: mid-March in 2022
Vegetative growth period: approx. 390 d
Initial flowering time: mid-April in 2022
Flowering period: approx. 27 d
Fruit maturation time: early-to-mid July in 2022
Fruit developing period: approx. 95 d
Fruit shape: short cylindrical
Single fruit weight: 581.6 g
Longitudinal diameter: 9.5 cm
Transverse diameter: 9.6 cm
Fruit shape index: 1.0
Immature fruit peel color: dark blackish green
Mature fruit peel color: light brown/ yellowish red
Fruit neck: absent
Fruit base: flat
Fruit top: flat
Fruitlets adhesion situation: separable
Fruit eyes appearance: convex/embossed
Fruit eyes depth: shallow
Fruit eyes arrangement: levorotation or dextrorotation
Fruit tumors: absent or many
Flesh color: pale yellow
Fruit aroma: faint/slight scent
Fruit flavour: strong sweet
Flesh texture: crisp/crunchy
Fruit appearance comprehensive evaluation: poor
Fruit quality comprehensive evaluation: excellent
Soluble solids content in flesh: 19.7%
Fruit ripening characteristics: late mature
Maturity consistency: nonuniform
Productivity: low

植株
Plant

现红
Open heart

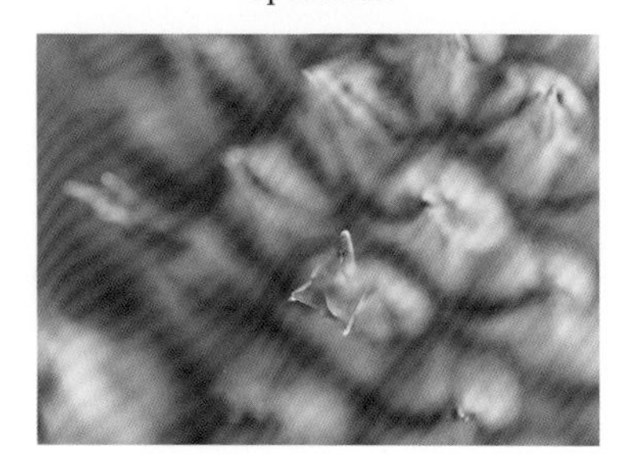

花
Flower

花序
Inflorescence

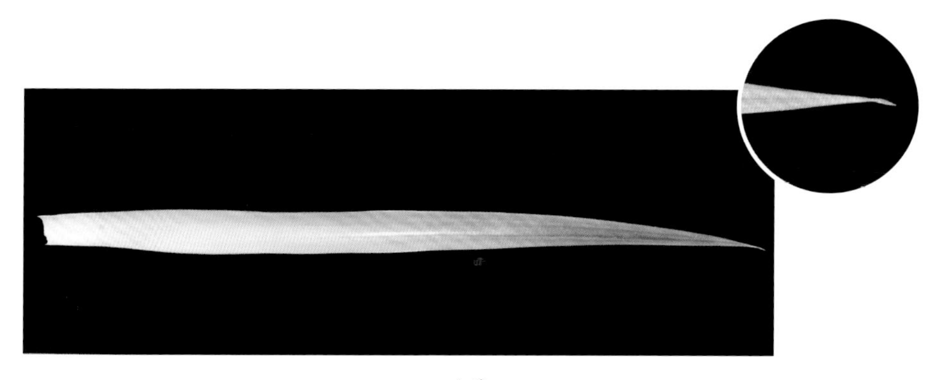

叶片
Leaf

带冠芽果
Fruit with crown bud

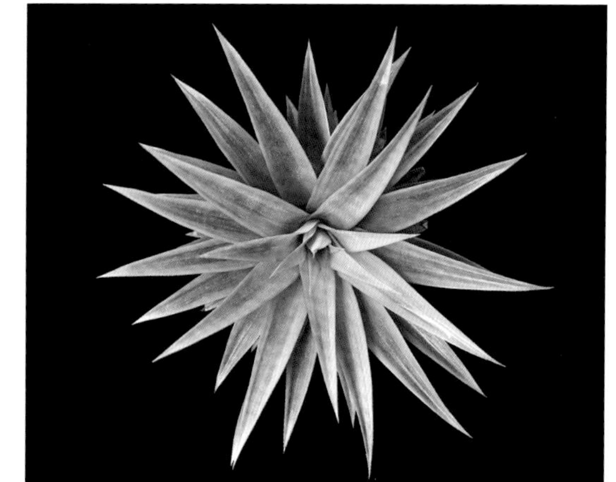

冠芽叶刺
Leaf spines in crown bud

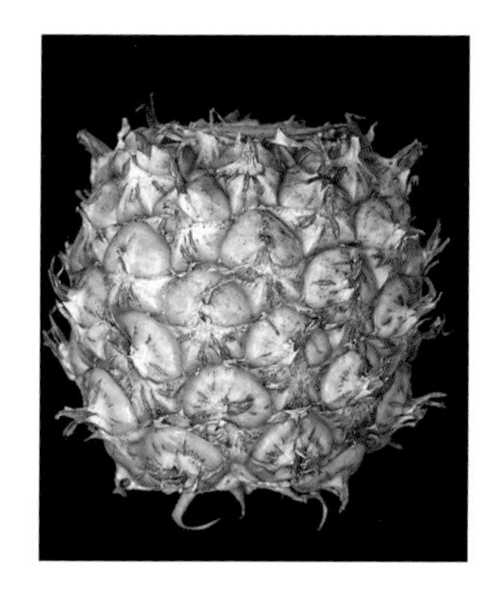

果实
Fruit

果实纵切
Fruit longitudinal section

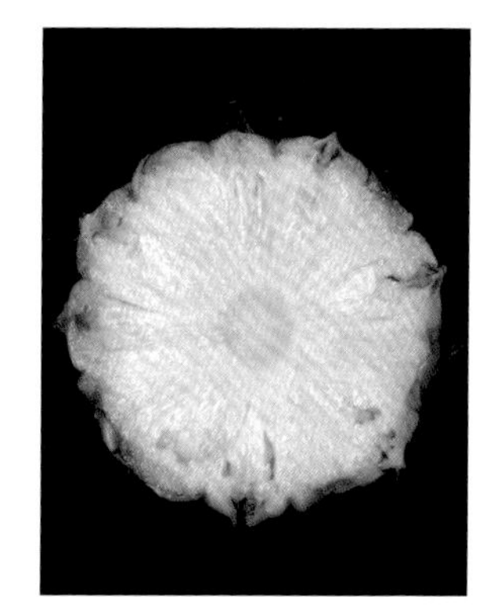

果实横切
Fruit transverse section

DN6

编号：ACCESSION000098
种质名称：DN6
原产地：越南
资源类型：引进品种
主要用途：鲜食或加工
种质来源地：2009 年中国热带农业科学院南亚热带作物研究所从越南引进
植株姿态：开张
定植期：2022 年 1 月上旬
现红期：2023 年 2 月下旬至 3 月上旬
营养生长期：约 425 d
初花期：2023 年 3 月下旬至 4 月上旬
花开放时间：约 19 d
成熟期：2023 年 7 月下旬
果实发育期：约 115 d
果实形状：圆筒形
单果质量：914.6 g
纵径：12.6 cm
横径：10.5 cm
果形指数：1.2
未成熟果实果皮颜色：暗绿色
成熟果实果皮颜色：亮黄 / 淡黄色
果颈：无
果基：平
果顶：钝圆
果实小果能否剥离：不可剥离
果眼外观：扁平或微凹
果眼深度：浅
果眼排列方式：左旋或右旋
果瘤：少或多
果肉颜色：淡黄色
果实香味：清香 / 微香
果实风味：清甜
果肉质地：滑
果实外观综合评价：好
果实品质综合评价：中
果肉可溶性固形物含量：13.8%
果实成熟特性：晚熟
成熟期的一致性：不一致
丰产性：中等

DN6

Code: ACCESSION000098
Name: DN6
Place of origin: Vietnam
Resource type: introduced variety
Main ways of consumption: fresh or processing
Initially introduction place: SSCRI, CATAS introduced from Vietnam in 2009
Plant posture: spreading
Planting time: early January in 2022
Open heart stage: late February to early March in 2023
Vegetative growth period: approx. 425 d
Initial flowering time: late March to early April in 2023
Flowering period: approx. 19 d
Fruit maturation time: late July in 2023
Fruit developing period: approx. 115 d
Fruit shape: cylindrical
Single fruit weight: 914.6 g
Longitudinal diameter: 12.6 cm
Transverse diameter: 10.5 cm
Fruit shape index: 1.2
Immature fruit peel color:dark green
Mature fruit peel color: bright yellow/light yellow
Fruit neck: absent
Fruit base: flat
Fruit top: bluntly round
Fruitlets adhesion situation: inseparable
Fruit eyes appearance: flat or slightly concave
Fruit eyes depth: shallow
Fruit eyes arrangement: levorotation or dextrorotation
Fruit tumors: few or many
Flesh color: pale yellow
Fruit aroma: faint/slight scent
Fruit flavour: mildly sweet
Flesh texture: smooth
Fruit appearance comprehensive evaluation: good
Fruit quality comprehensive evaluation: medium
Soluble solids content in flesh: 13.8%
Fruit ripening characteristics: late mature
Maturity consistency: nonuniform
Productivity:normal level

植株
Plant

现红
Open heart

花
Flower

花序
Inflorescence

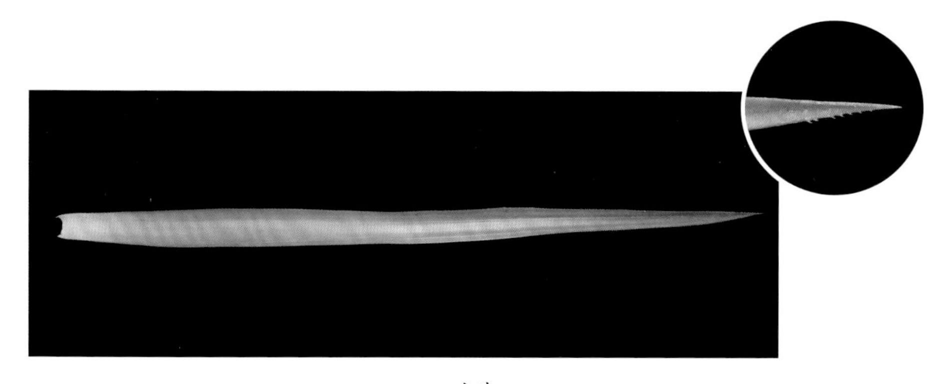

叶片
Leaf

带冠芽果
Fruit with crown bud

冠芽叶刺
Leaf spines in crown bud

果实
Fruit

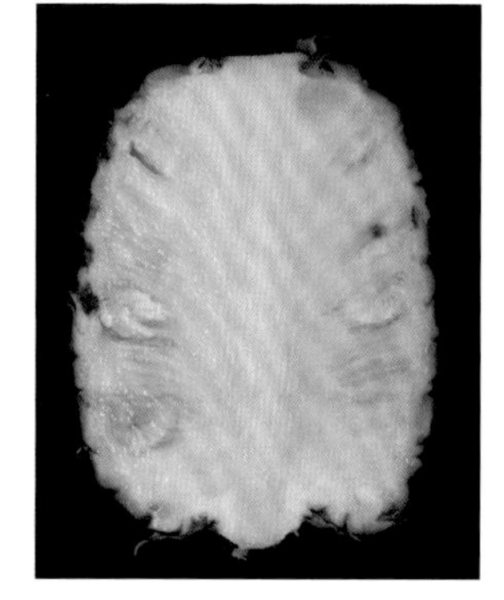

果实纵切
Fruit longitudinal section

果实横切
Fruit transverse section

越南引皇后 2 号

编号：ACCESSION000063
种质名称：越南引皇后 2 号
原产地：越南
资源类型：引进品种
主要用途：鲜食或加工
种质来源地：2009 年中国热带农业科学院南亚热带作物研究所从越南引进
植株姿态：开张
定植期：2021 年 12 月下旬
现红期：2023 年 2 月下旬
营养生长期：约 425 d
初花期：2023 年 3 月下旬
花开放时间：约 20 d
成熟期：2023 年 6 月中旬
果实发育期：约 80 d
果实形状：长圆筒形
单果质量：820.6 g
纵径：14.1 cm
横径：10.2 cm
果形指数：1.4
未成熟果实果皮颜色：淡绿色
成熟果实果皮颜色：金黄 / 鲜黄色
果颈：无
果基：平
果顶：平
果实小果能否剥离：可剥离
果眼外观：突起 / 隆起
果眼深度：深
果眼排列方式：右旋
果瘤：少或多
果肉颜色：淡黄色
果实香味：清香 / 微香
果实风味：酸甜
果肉质地：滑
果实外观综合评价：优
果实品质综合评价：好
果肉可溶性固形物含量：17.8%
果实成熟特性：中熟
成熟期的一致性：一致
丰产性：一般

Queen No. 2 from Vietnam

Code: ACCESSION000063
Name: Queen No. 2 from Vietnam
Place of origin: Vietnam
Resource type: introduced variety
Main ways of consumption: fresh or processing
Initially introduction place: SSCRI, CATAS introduced from Vietnam in 2009
Plant posture: spreading
Planting time: late December in 2021
Open heart stage: late February in 2023
Vegetative growth period: approx. 425 d
Initial flowering time: late March in 2023
Flowering period: approx. 20 d
Fruit maturation time: mid-June in 2023
Fruit developing period: approx. 80 d
Fruit shape: long cylindrical
Single fruit weight: 820.6 g
Longitudinal diameter: 14.1 cm
Transverse diameter: 10.2 cm
Fruit shape index: 1.4
Immature fruit peel color: light green
Mature fruit peel color: golden yellow/vivid yellow
Fruit neck: absent
Fruit base: flat
Fruit top: flat
Fruitlets adhesion situation: separable
Fruit eyes appearance: convex/embossed
Fruit eyes depth: deep
Fruit eyes arrangement: dextrorotation
Fruit tumors: few or many
Flesh color: pale yellow
Fruit aroma: faint/slight scent
Fruit flavour: sour-sweet
Flesh texture: smooth
Fruit appearance comprehensive evaluation: excellent
Fruit quality comprehensive evaluation: good
Soluble solids content in flesh: 17.8%
Fruit ripening characteristics: medium mature
Maturity consistency: uniform
Productivity: normal level

植株
Plant

现红
Open heart

花
Flower

花序
Inflorescence

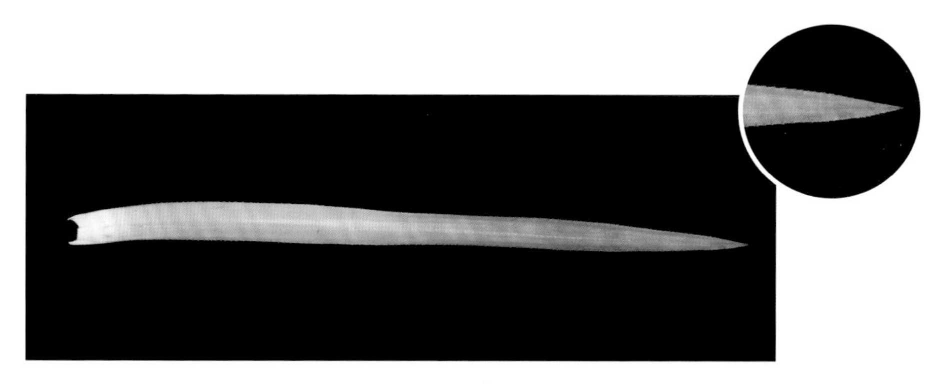

叶片
Leaf

带冠芽果
Fruit with crown bud

冠芽叶刺
Leaf spines in crown bud

果实
Fruit

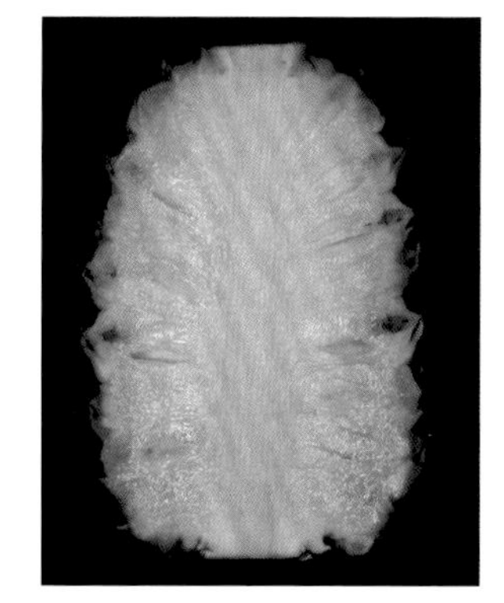

果实纵切
Fruit longitudinal section

果实横切
Fruit transverse section

越南引皇后 3 号

编号：ACCESSION000085
种质名称：越南引皇后 3 号
原产地：越南
资源类型：引进品种
主要用途：鲜食或加工
种质来源地：2010 年中国热带农业科学院南亚热带作物研究所从越南引进
植株姿态：开张
定植期：2022 年 1 月上旬
现红期：2023 年 2 月中旬
营养生长期：约 420 d
初花期：2023 年 3 月中旬
花开放时间：约 18 d
成熟期：2023 年 6 月中旬
果实发育期：约 90 d
果实形状：圆筒形
单果质量：658.1 g
纵径：12.6 cm
横径：9.7 cm
果形指数：1.3
未成熟果实果皮颜色：淡绿色
成熟果实果皮颜色：金黄 / 鲜黄色
果颈：有
果基：突起
果顶：平
果实小果能否剥离：可剥离
果眼外观：突起 / 隆起
果眼深度：较浅
果眼排列方式：右旋
果瘤：少
果肉颜色：黄色
果实香味：清香 / 微香
果实风味：酸甜
果肉质地：滑
果实外观综合评价：好
果实品质综合评价：中
果肉可溶性固形物含量：14.5%
果实成熟特性：中熟
成熟期的一致性：一致
丰产性：低产

Queen No. 3 from Vietnam

Code: ACCESSION000085
Name: Queen No. 3 from Vietnam
Place of origin: Vietnam
Resource type: introduced variety
Main ways of consumption: fresh or processing
Initially introduction place: SSCRI, CATAS introduced from Vietnam in 2010
Plant posture: spreading
Planting time: early January in 2022
Open heart stage: mid-February in 2023
Vegetative growth period: approx. 420 d
Initial flowering time: mid-March in 2023
Flowering period: approx. 18 d
Fruit maturation time: mid-June in 2023
Fruit developing period: approx. 90 d
Fruit shape: cylindrical
Single fruit weight: 658.1 g
Longitudinal diameter: 12.6 cm
Transverse diameter: 9.7 cm
Fruit shape index: 1.3
Immature fruit peel color: light green
Mature fruit peel color: golden yellow/vivid yellow
Fruit neck: present
Fruit base: protrusive
Fruit top: flat
Fruitlets adhesion situation: separable
Fruit eyes appearance: convex/embossed
Fruit eyes depth: relativele shallow
Fruit eyes arrangement: dextrorotation
Fruit tumors: few
Flesh color: yellow
Fruit aroma: faint/slight scent
Fruit flavour: sour-sweet
Flesh texture: smooth
Fruit appearance comprehensive evaluation: good
Fruit quality comprehensive evaluation: medium
Soluble solids content in flesh: 14.5%
Fruit ripening characteristics: medium mature
Maturity consistency: uniform
Productivity: low

植株
Plant

现红
Open heart

花
Flower

花序
Inflorescence

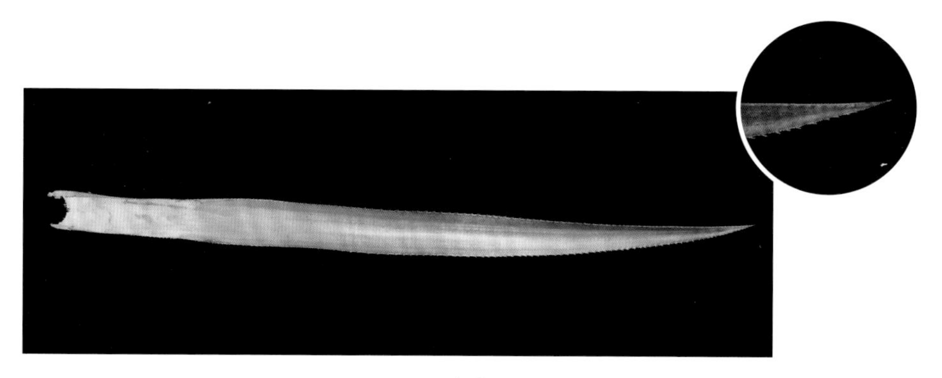

叶片
Leaf

带冠芽果
Fruit with crown bud

冠芽叶刺
Leaf spines in crown bud

果实
Fruit

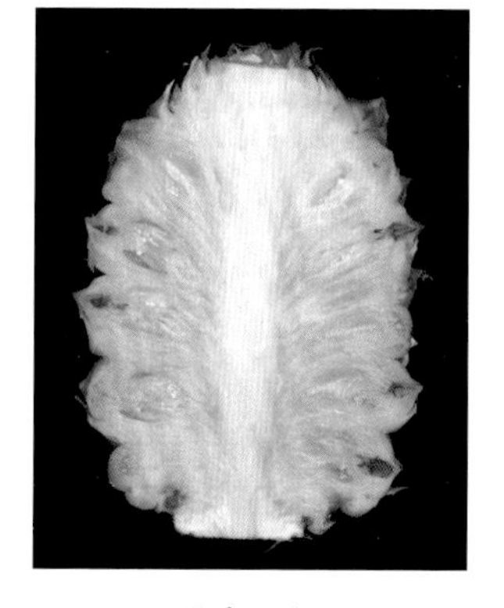

果实纵切
Fruit longitudinal section

果实横切
Fruit transverse section

越南引无刺卡因 1 号

编号：ACCESSION000041
种质名称：越南引无刺卡因 1 号
原产地：越南
资源类型：引进品种
主要用途：鲜食或加工
种质来源地：2008 年中国热带农业科学院南亚热带作物研究所从越南引进
植株姿态：开张
定植期：2022 年 1 月上旬
现红期：2023 年 2 月下旬至 3 月上旬
营养生长期：约 425 d
初花期：2023 年 3 月下旬至 4 月上旬
花开放时间：约 19 d
成熟期：2023 年 7 月下旬
果实发育期：约 115 d
果实形状：圆筒形
单果质量：964.2 g
纵径：13.0 cm
横径：11.1 cm
果形指数：1.2
未成熟果实果皮颜色：暗墨绿色
成熟果实果皮颜色：暗黄 / 深黄色
果颈：无
果基：平
果顶：平
果实小果能否剥离：不可剥离
果眼外观：微隆起
果眼深度：浅
果眼排列方式：左旋
果瘤：少
果肉颜色：淡黄色
果实香味：清香 / 微香
果实风味：酸甜
果肉质地：粗糙
果实外观综合评价：中
果实品质综合评价：中
果肉可溶性固形物含量：15.2%
果实成熟特性：晚熟
成熟期的一致性：不一致
丰产性：一般

Smooth Cayenne No. 1 from Vietnam

Code: ACCESSION000041
Name: Smooth Cayenne No. 1 from Vietnam
Place of origin: Vietnam
Resource type: introduced variety
Main ways of consumption: fresh or processing
Initially introduction place: SSCRI, CATAS introduced from Vietnam in 2008
Plant posture: spreading
Planting time: early January in 2022
Open heart stage: late February to early March in 2023
Vegetative growth period: approx. 425 d
Initial flowering time: late March to early April in 2023
Flowering period: approx. 19 d
Fruit maturation time: late July in 2023
Fruit developing period: approx. 115 d
Fruit shape: cylindrical
Single fruit weight: 964.2 g
Longitudinal diameter: 13.0 cm
Transverse diameter: 11.1 cm
Fruit shape index: 1.2
Immature fruit peel color: dark blackish green
Mature fruit peel color: dark/deep yellow
Fruit neck: absent
Fruit base: flat
Fruit top: flat
Fruitlets adhesion situation: inseparable
Fruit eyes appearance: slightly embossed
Fruit eyes depth: shallow
Fruit eyes arrangement: levorotation
Fruit tumors: few
Flesh color: pake yellow
Fruit aroma: faint/slight scent
Fruit flavour: sour-sweet
Flesh texture: coarse
Fruit appearance comprehensive evaluation: medium
Fruit quality comprehensive evaluation: medium
Soluble solids content in flesh: 15.2%
Fruit ripening characteristics: late mature
Maturity consistency: nonuniform
Productivity: normal level

植株
Plant

现红
Open heart

花
Flower

花序
Inflorescence

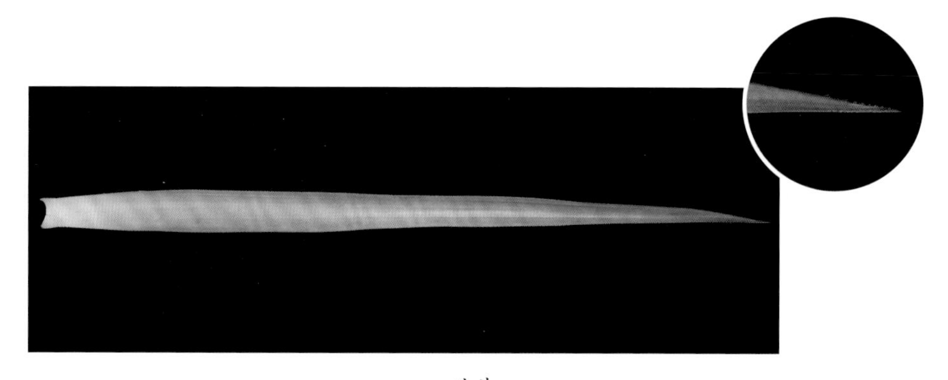

叶片
Leaf

带冠芽果
Fruit with crown bud

冠芽叶刺
Leaf spines in crown bud

果实
Fruit

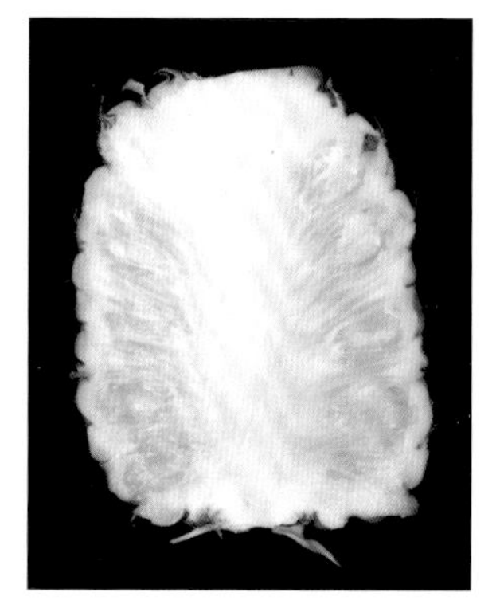

果实纵切
Fruit longitudinal section

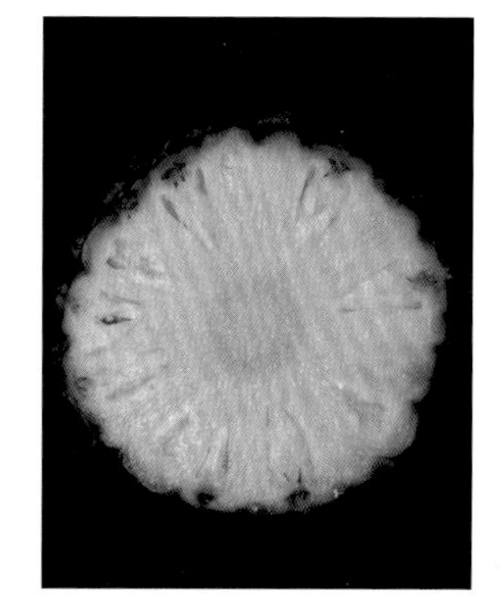

果实横切
Fruit transverse section

Moroshious

编号：ACCESSION000015
种质名称：Moroshious
原产地：印度
资源类型：引进品种
主要用途：鲜食或加工
种质来源地：2008 年中国热带农业科学院南亚热带作物研究所从印度引进
植株姿态：开张
定植期：2022 年 1 月上旬
现红期：2023 年 2 月下旬至 3 月上旬
营养生长期：约 425 d
初花期：2023 年 3 月下旬
花开放时间：约 18 d
成熟期：2023 年 6 月中旬
果实发育期：约 80 d
果实形状：圆筒形
单果质量：741.4 g
纵径：12.5 cm
横径：10.2 cm
果形指数：1.2
未成熟果实果皮颜色：绿色
成熟果实果皮颜色：亮黄 / 淡黄色
果颈：无
果基：平
果顶：平
果实小果能否剥离：可剥离
果眼外观：突起 / 隆起
果眼深度：较深
果眼排列方式：左旋
果瘤：无或少
果肉颜色：黄色
果实香味：清香 / 微香
果实风味：酸甜
果肉质地：脆 / 爽脆
果实外观综合评价：中
果实品质综合评价：好
果肉可溶性固形物含量：19.3%
果实成熟特性：晚熟
成熟期的一致性：一致
丰产性：一般

Moroshious

Code: ACCESSION000015
Name: Moroshious
Place of origin: India
Resource type: introduced variety
Main ways of consumption: fresh or processing
Initially introduction place: SSCRI, CATAS introduced from India in 2008
Plant posture: spreading
Planting time: early January in 2022
Open heart stage: late February to early March in 2023
Vegetative growth period: approx. 425 d
Initial flowering time: late March in 2023
Flowering period: approx. 18 d
Fruit maturation time: mid-June in 2023
Fruit developing period: approx. 80 d
Fruit shape: cylindrical
Single fruit weight: 741.4 g
Longitudinal diameter: 12.5 cm
Transverse diameter: 10.2 cm
Fruit shape index: 1.2
Immature fruit peel color: green
Mature fruit peel color: bright yellow/light yellow
Fruit neck: absent
Fruit base: flat
Fruit top: flat
Fruitlets adhesion situation: separable
Fruit eyes appearance: convex/embossed
Fruit eyes depth: relatively deep
Fruit eyes arrangement: levorotation
Fruit tumors: absent or few
Flesh color: yellow
Fruit aroma: faint/slight scent
Fruit flavour: sour-sweet
Flesh texture: crisp/crunchy
Fruit appearance comprehensive evaluation: medium
Fruit quality comprehensive evaluation: good
Soluble solids content in flesh: 19.3%
Fruit ripening characteristics: late mature
Maturity consistency: uniform
Productivity: normal level

植株
Plant

现红
Open heart

花
Flower

花序
Inflorescence

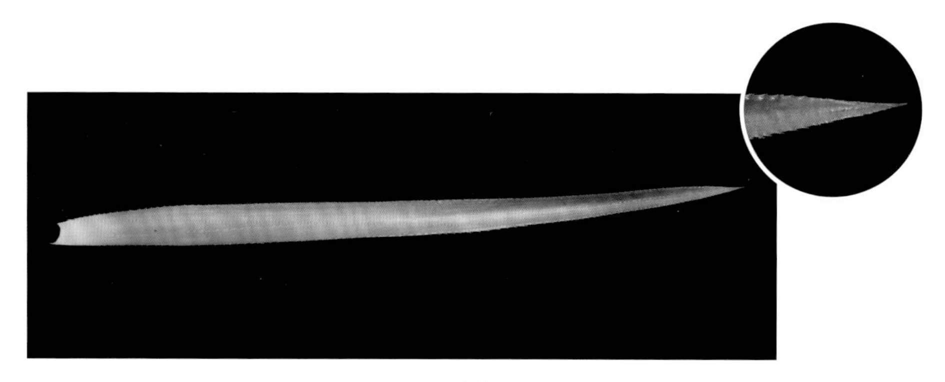

叶片
Leaf

带冠芽果
Fruit with crown bud

冠芽叶刺
Leaf spines in crown bud

果实
Fruit

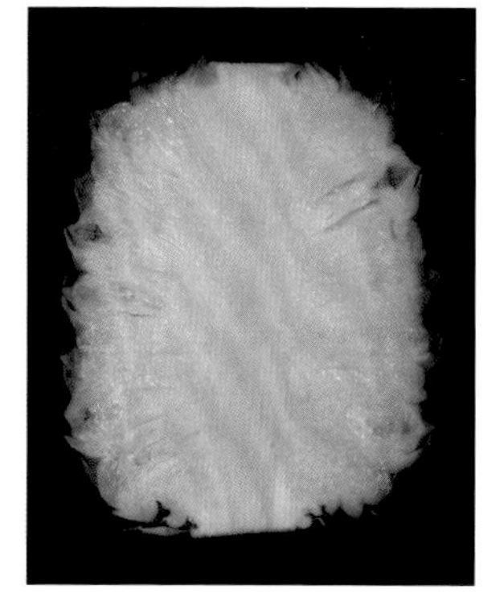

果实纵切
Fruit longitudinal section

果实横切
Fruit transverse section

印尼无刺

编号：ACCESSION000070
种质名称：印尼无刺
原产地：印度尼西亚
资源类型：引进品种
主要用途：鲜食或加工
种质来源地：2009 年中国热带农业科学院南亚热带作物研究所从印度尼西亚引进
植株姿态：开张
定植期：2021 年 12 月下旬
现红期：2023 年 2 月下旬至 3 月上旬
营养生长期：约 430 d
初花期：2023 年 4 月上旬
花开放时间：约 25 d
成熟期：2023 年 7 月中旬
果实发育期：约 100 d
果实形状：短圆筒形
单果质量：1050.1 g
纵径：13.6 cm
横径：11.0 cm
果形指数：1.2
未成熟果实果皮颜色：绿色
成熟果实果皮颜色：黄色，带绿斑
果颈：无
果基：平
果顶：平
果实小果能否剥离：不可剥离
果眼外观：扁平或微凹
果眼深度：较浅
果眼排列方式：右旋
果瘤：无
果肉颜色：淡黄色
果实香味：清香 / 微香
果实风味：酸甜
果肉质地：脆 / 爽脆
果实外观综合评价：差
果实品质综合评价：中
果肉可溶性固形物含量：13.6%
果实成熟特性：晚熟
成熟期的一致性：一致
丰产性：一般

Indonesia Wuci

Code: ACCESSION000070
Name: Indonesia Wuci
Place of origin: Indonesia
Resource type: introduced variety
Main ways of consumption: fresh or processing
Initially introduction place: SSCRI, CATAS introduced from Indonesia in 2009
Plant posture: spreading
Planting time: late December in 2021
Open heart stage: late February to early March in 2023
Vegetative growth period: approx. 430 d
Initial flowering time: early April in 2023
Flowering period: approx. 25 d
Fruit maturation time: mid-July in 2023
Fruit developing period: approx. 100 d
Fruit shape: short cylindrical
Single fruit weight: 1050.1 g
Longitudinal diameter: 13.6 cm
Transverse diameter: 11.0 cm
Fruit shape index: 1.2
Immature fruit peel color: green
Mature fruit peel color: yellow with green stripe
Fruit neck: absent
Fruit base: flat
Fruit top: flat
Fruitlets adhesion situation: inseparable
Fruit eyes appearance: flat or slightly concave
Fruit eyes depth: relatively shallow
Fruit eyes arrangement: dextrorotation
Fruit tumors: absent
Flesh color: pale yellow
Fruit aroma: faint/slight scent
Fruit flavour: sour-sweet
Flesh texture: crisp/crunchy
Fruit appearance comprehensive evaluation: poor
Fruit quality comprehensive evaluation: medium
Soluble solids content in flesh: 13.6%
Fruit ripening characteristics: late mature
Maturity consistency: uniform
Productivity: normal level

植株
Plant

现红
Open heart

花
Flower

花序
Inflorescence

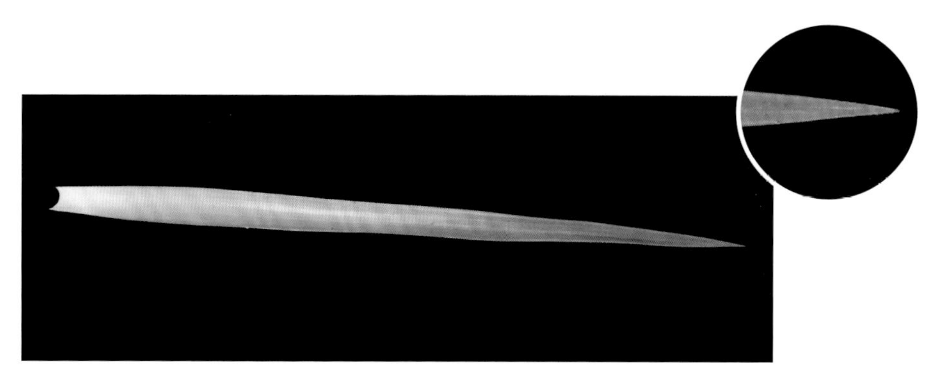

叶片
Leaf

带冠芽果
Fruit with crown bud

冠芽叶刺
Leaf spines in crown bud

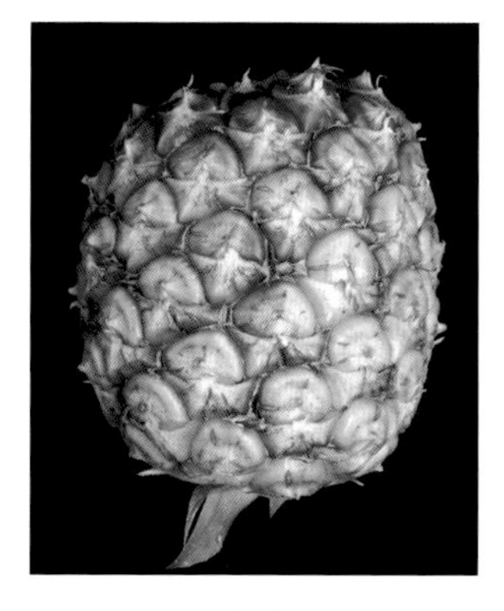

果实
Fruit

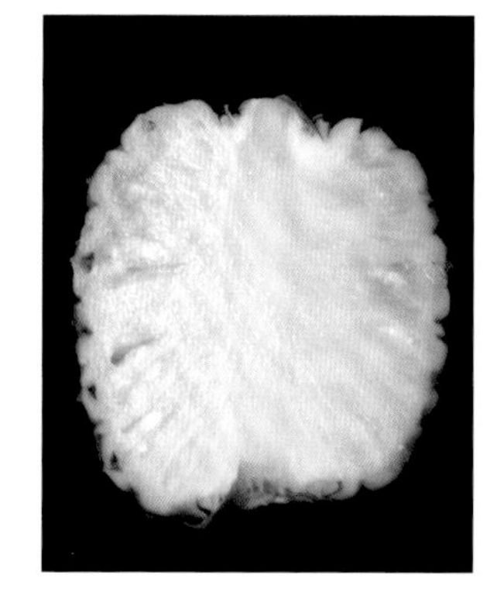

果实纵切
Fruit longitudinal section

果实横切
Fruit transverse section

Nenas Sinadu

编号：ACCESSION000057
种质名称：Nenas Sinadu
原产地：印度尼西亚
资源类型：引进品种
主要用途：鲜食或加工
种质来源地：2008 年中国热带农业科学院南亚热带作物研究所从印度尼西亚引进
植株姿态：开张
定植期：2022 年 1 月上旬
现红期：2023 年 3 月上旬
营养生长期：约 425 d
初花期：2023 年 4 月上旬
花开放时间：约 25 d
成熟期：2023 年 7 月上旬
果实发育期：约 90 d
果实形状：圆筒形
单果质量：790.4 g
纵径：11.4 cm
横径：10.2 cm
果形指数：1.1
未成熟果实果皮颜色：暗墨绿色
成熟果实果皮颜色：浅褐 / 黄红
果颈：无
果基：平
果顶：平
果实小果能否剥离：不可剥离
果眼外观：扁平或微凹
果眼深度：浅
果眼排列方式：左旋
果瘤：少
果肉颜色：淡黄色
果实香味：清香 / 微香
果实风味：酸甜
果肉质地：粗糙
果实外观综合评价：中
果实品质综合评价：中
果肉可溶性固形物含量：14.9%
果实成熟特性：中熟
成熟期的一致性：不一致
丰产性：低产

Nenas Sinadu

Code: ACCESSION000057
Name: Nenas Sinadu
Place of origin: Indonesia
Resource type: introduced variety
Main ways of consumption: fresh or processing
Initially introduction place: SSCRI, CATAS introduced from Indonesia in 2008
Plant posture: spreading
Planting time: early January in 2022
Open heart stage: early March in 2023
Vegetative growth period: approx. 425 d
Initial flowering time: early April in 2023
Flowering period: approx. 25 d
Fruit maturation time: early July in 2023
Fruit developing period: approx. 90 d
Fruit shape: cylindrical
Single fruit weight: 790.4 g
Longitudinal diameter: 11.4 cm
Transverse diameter: 10.2 cm
Fruit shape index: 1.1
Immature fruit peel color: dark blackish green
Mature fruit peel color: light brown/ yellowish red
Fruit neck: absent
Fruit base: flat
Fruit top: flat
Fruitlets adhesion situation: inseparable
Fruit eyes appearance: flat or slightly concave
Fruit eyes depth: shallow
Fruit eyes arrangement: levorotation
Fruit tumors: few
Flesh color: pale yellow
Fruit aroma: faint/slight scent
Fruit flavour: sour-sweet
Flesh texture: coarse
Fruit appearance comprehensive evaluation: medium
Fruit quality comprehensive evaluation: medium
Soluble solids content in flesh: 14.9%
Fruit ripening characteristics: medium mature
Maturity consistency: nonuniform
Productivity: low

植株
Plant

现红
Open heart

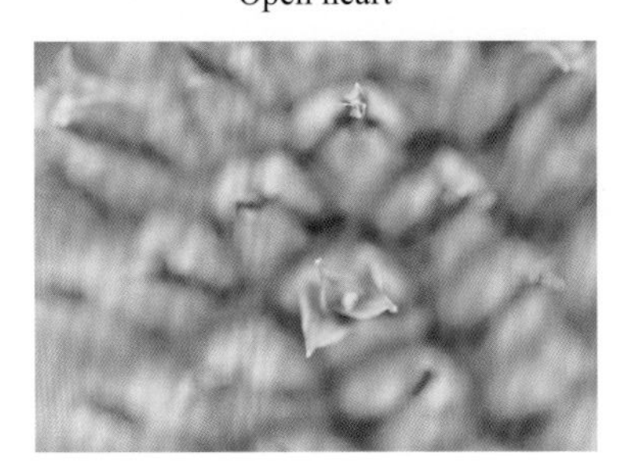

花
Flower

花序
Inflorescence

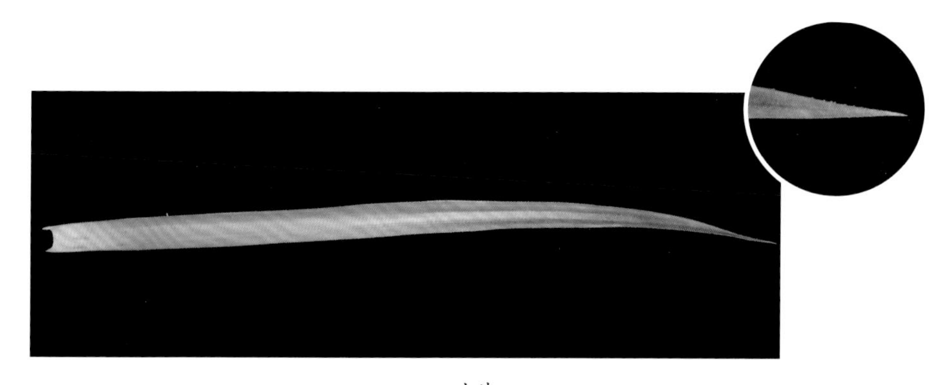

叶片
Leaf

带冠芽果
Fruit with crown bud

冠芽叶刺
Leaf spines in crown bud

果实
Fruit

果实纵切
Fruit longitudinal section

果实横切
Fruit transverse section

Nenas Cayenne × Nenas Bogor

编号：ACCESSION000059
种质名称：Nenas Cayenne × Nenas Bogor
原产地：印度尼西亚
资源类型：引进品种
主要用途：鲜食或加工
种质来源地：2008年中国热带农业科学院南亚热带作物研究所从印度尼西亚引进
植株姿态：开张
定植期：2022年1月上旬
现红期：2023年2月下旬
营养生长期：约420 d
初花期：2023年3月下旬
花开放时间：约20 d
成熟期：2023年6月中下旬
果实发育期：约85 d
果实形状：圆筒形
单果质量：522.9 g
纵径：10.8 cm
横径：9.4 cm
果形指数：1.1
未成熟果实果皮颜色：淡绿色
成熟果实果皮颜色：黄色，带绿斑
果颈：无
果基：平
果顶：平
果实小果能否剥离：可剥离
果眼外观：突起 / 隆起
果眼深度：较浅
果眼排列方式：右旋
果瘤：无
果肉颜色：淡黄色
果实香味：清香 / 微香
果实风味：甜酸
果肉质地：滑
果实外观综合评价：优
果实品质综合评价：好
果肉可溶性固形物含量：18.2%
果实成熟特性：中熟
成熟期的一致性：不一致
丰产性：低产

Nenas Cayenne×Nenas Bogor

Code: ACCESSION000059
Name: Nenas Cayenne × Nenas Bogor
Place of origin: Indonesia
Resource type: introduced variety
Main ways of consumption: fresh or processing
Initially introduction place: SSCRI, CATAS introduced from Indonesia in 2008
Plant posture: spreading
Planting time: early January in 2022
Open heart stage: late February in 2023
Vegetative growth period: approx. 420 d
Initial flowering time: late March in 2023
Flowering period: approx. 20 d
Fruit maturation time: mid-to-late June in 2023
Fruit developing period: approx. 85 d
Fruit shape: cylindrical
Single fruit weight: 522.9 g
Longitudinal diameter: 10.8 cm
Transverse diameter: 9.4 cm
Fruit shape index: 1.1
Immature fruit peel color: light green
Mature fruit peel color: yellow with green stripe
Fruit neck: absent
Fruit base: flat
Fruit top: flat
Fruitlets adhesion situation: separable
Fruit eyes appearance: convex/embossed
Fruit eyes depth: relatively shallow
Fruit eyes arrangement: dextrorotation
Fruit tumors: absent
Flesh color: pale yellow
Fruit aroma: faint/slight scent
Fruit flavour: sweet-sour
Flesh texture: smooth
Fruit appearance comprehensive evaluation: excellent
Fruit quality comprehensive evaluation: good
Soluble solids content in flesh: 18.2%
Fruit ripening characteristics: medium mature
Maturity consistency: nonuniform
Productivity: low

植株
Plant

现红
Open heart

花
Flower

花序
Inflorescence

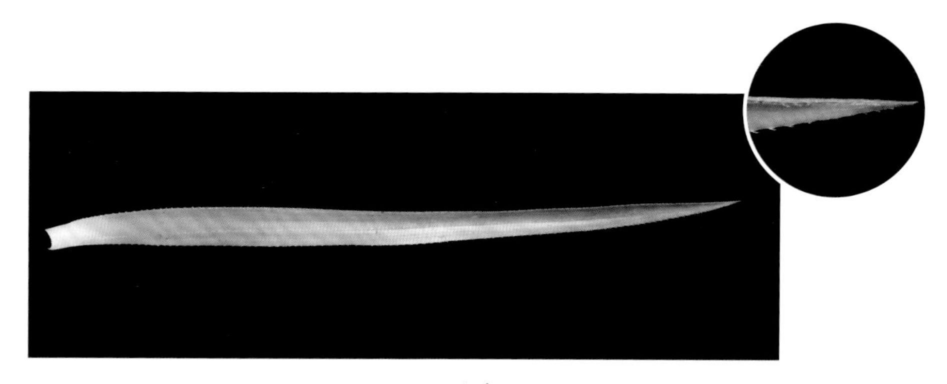

叶片
Leaf

带冠芽果
Fruit with crown bud

冠芽叶刺
Leaf spines in crown bud

果实
Fruit

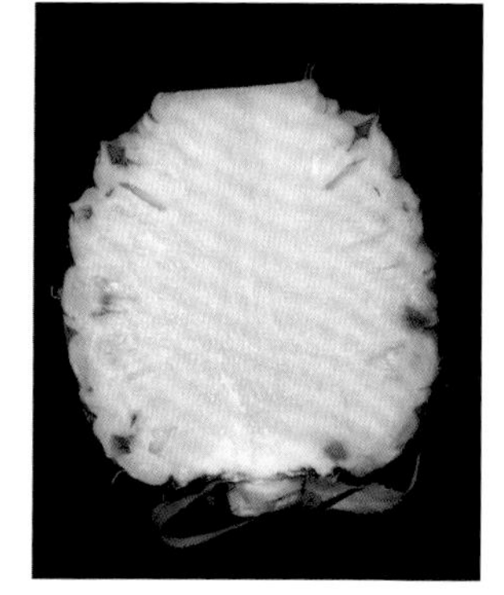

果实纵切
Fruit longitudinal section

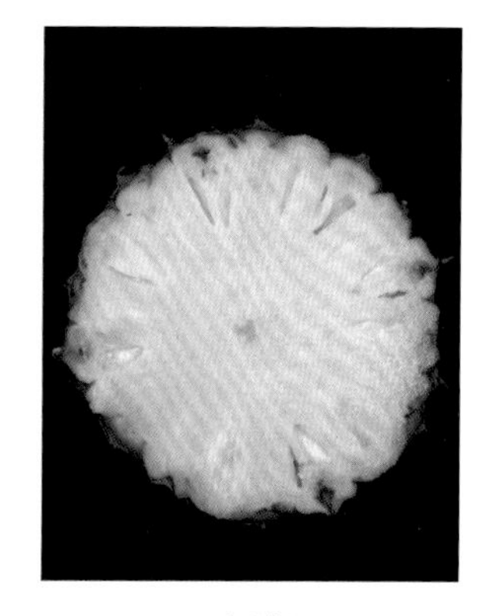

果实横切
Fruit transverse section

杂 2

编号：ACCESSION000080
种质名称：杂 2
原产地：马来西亚
资源类型：引进品种
主要用途：鲜食或加工
种质来源地：2008 年中国热带农业科学院南亚热带作物研究所从马来西亚引进
植株姿态：开张
定植期：2021 年 12 月下旬
现红期：2023 年 3 月中旬
营养生长期：约 440 d
初花期：2023 年 4 月上中旬
花开放时间：约 17 d
成熟期：2023 年 7 月下旬
果实发育期：约 105 d
果实形状：圆锥形
单果质量：1103.6 g
纵径：14.0 cm
横径：11.3 cm
果形指数：1.2
未成熟果实果皮颜色：暗墨绿色
成熟果实果皮颜色：亮黄色
果颈：有
果基：平
果顶：钝圆
果实小果能否剥离：不可剥离
果眼外观：微隆起
果眼深度：浅
果眼排列方式：左旋
果瘤：无
果肉颜色：黄色
果实香味：无
果实风味：甜酸
果肉质地：滑
果实外观综合评价：中
果实品质综合评价：中
果肉可溶性固形物含量：12.4%
果实成熟特性：晚熟
成熟期的一致性：一致
丰产性：丰产

Hybrid No. 2

Code: ACCESSION000080
Name: Hybrid No. 2
Resource type: introduced variety
Main ways of consumption: fresh or processing
Initially introduction place: SSCRI, CATAS introduced from Malaysia in 2008
Plant posture: spreading
Planting time: late December in 2021
Open heart stage: mid-March in 2023
Vegetative growth period: approx. 440 d
Initial flowering time: early-to-mid April in 2023
Flowering period: approx. 17 d
Fruit maturation time: late July in 2023
Fruit developing period: approx. 105 d
Fruit shape: conical
Single fruit weight: 1103.6 g
Longitudinal diameter: 14.0 cm
Transverse diameter: 11.3 cm
Fruit shape index: 1.2
Immature fruit peel color: dark blackish green
Mature fruit peel color: bright yellow
Fruit neck: present
Fruit base: flat
Fruit top: bluntly round
Fruitlets adhesion situation: inseparable
Fruit eyes appearance: slightly embossed
Fruit eyes depth: shallow
Fruit eyes arrangement: levorotation
Fruit tumors: absent
Flesh color: yellow
Fruit aroma: no scent
Fruit flavour: sweet-sour
Flesh texture: smooth
Fruit appearance comprehensive evaluation: medium
Fruit quality comprehensive evaluation: medium
Soluble solids content in flesh: 12.4%
Fruit ripening characteristics: late mature
Maturity consistency: uniform
Productivity: productive

植株
Plant

现红
Open heart

花
Flower

花序
Inflorescence

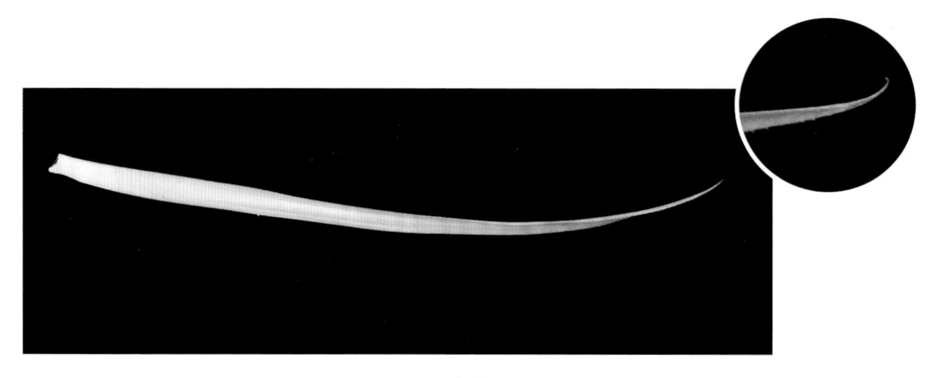

叶片
Leaf

带冠芽果
Fruit with crown bud

冠芽叶刺
Leaf spines in crown bud

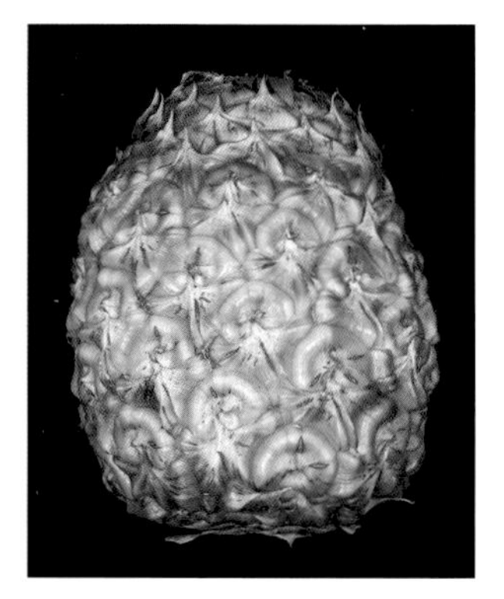

果实
Fruit

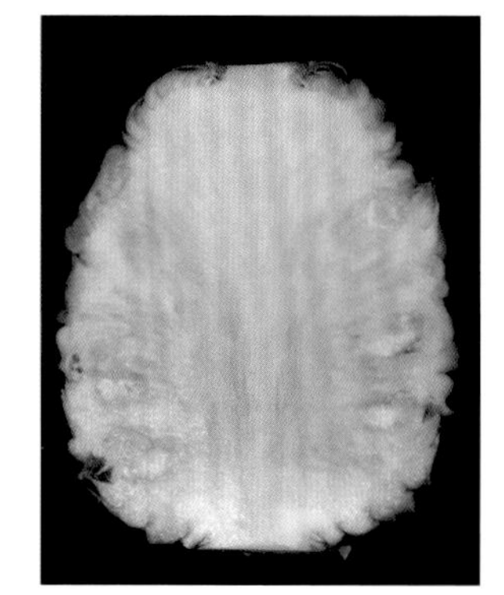

果实纵切
Fruit longitudinal section

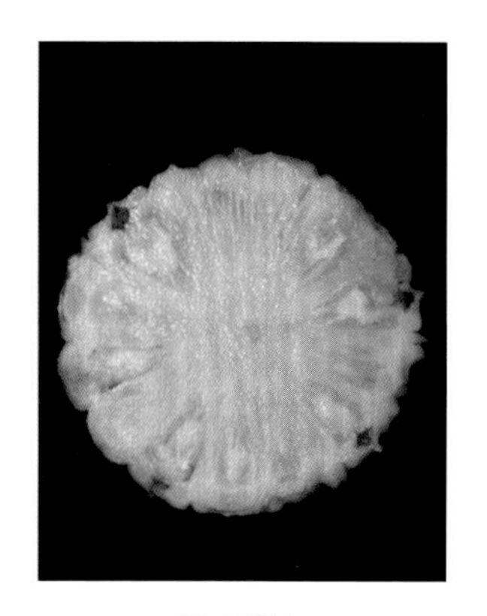

果实横切
Fruit transverse section

Moris

编号：ACCESSION000051
种质名称：Moris
原产地：马来西亚
资源类型：引进品种
主要用途：鲜食或加工
种质来源地：2008 年中国热带农业科学院南亚热带作物研究所从马来西亚引进
植株姿态：开张
定植期：2021 年 2 月下旬
现红期：2022 年 3 月上旬
营养生长期：约 375 d
初花期：2022 年 3 月下旬至 4 月上旬
花开放时间：约 14 d
成熟期：2022 年 7 月上旬
果实发育期：约 90 d
果实形状：圆筒形
单果质量：343.0 g
纵径：9.1 cm
横径：8.3 cm
果形指数：1.1
未成熟果实果皮颜色：暗绿色
成熟果实果皮颜色：亮黄 / 淡黄色
果颈：无
果基：平
果顶：平
果实小果能否剥离：可剥离
果眼外观：突起 / 隆起
果眼深度：较浅
果眼排列方式：右旋
果瘤：无
果肉颜色：黄色
果实香味：清香 / 微香
果实风味：甜酸
果肉质地：脆 / 爽脆
果实外观综合评价：中
果实品质综合评价：中
果肉可溶性固形物含量：16.1%
果实成熟特性：中熟
成熟期的一致性：一致
丰产性：低产

Moris

Code: ACCESSION000051
Name: Moris
Place of origin: Malaysia
Resource type: introduced variety
Main ways of consumption: fresh or processing
Initially introduction place: SSCRI, CATAS introduced from Malaysia in 2008
Plant posture: spreading
Planting time: late February in 2021
Open heart stage: early March in 2022
Vegetative growth period: approx. 375 d
Initial flowering time: late March to early April in 2022
Flowering period: approx. 14 d
Fruit maturation time: early July in 2022
Fruit developing period: approx. 90 d
Fruit shape: cylindrical
Single fruit weight: 343.0 g
Longitudinal diameter: 9.1 cm
Transverse diameter: 8.3 cm
Fruit shape index: 1.1
Immature fruit peel color: dark green
Mature fruit peel color: bright yellow/ light yellow
Fruit neck: absent
Fruit base: flat
Fruit top: flat
Fruitlets adhesion situation: separable
Fruit eyes appearance: convex/embossed
Fruit eyes depth: relatively shallow
Fruit eyes arrangement: dextrorotation
Fruit tumors: absent
Flesh color: yellow
Fruit aroma: faint/slight scent
Fruit flavour: sweet-sour
Flesh texture: crisp/crunchy
Fruit appearance comprehensive evaluation: medium
Fruit quality comprehensive evaluation: medium
Soluble solids content in flesh: 16.1%
Fruit ripening characteristics: medium mature
Maturity consistency: uniform
Productivity: low

植株
Plant

现红
Open heart

花
Flower

花序
Inflorescence

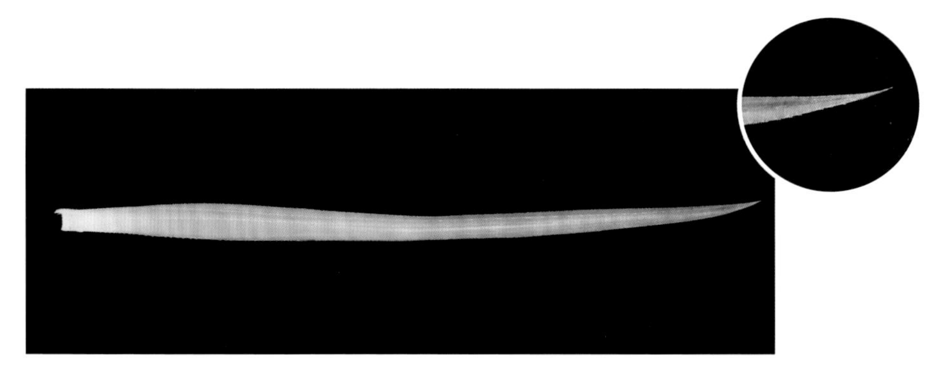

叶片
Leaf

带冠芽果
Fruit with crown bud

冠芽叶刺
Leaf spines in crown bud

果实
Fruit

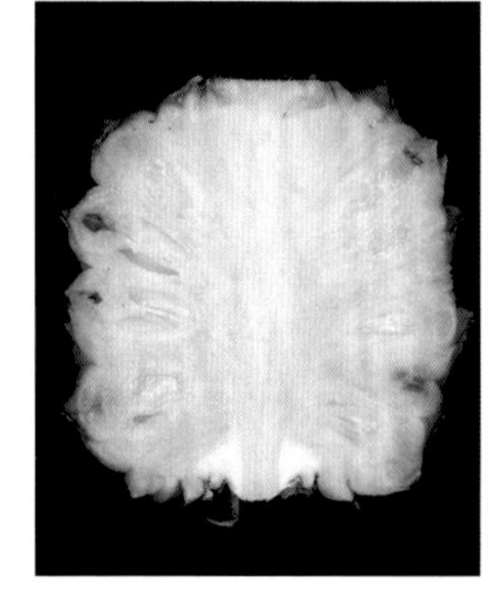

果实纵切
Fruit longitudinal section

果实横切
Fruit transverse section

Moris Gajah

编号：ACCESSION000077
种质名称：Moris Gajah
原产地：马来西亚
资源类型：引进品种
主要用途：鲜食或加工
种质来源地：2010 年中国热带农业科学院南亚热带作物研究所从马来西亚引进
植株姿态：开张
定植期：2022 年 1 月上旬
现红期：2023 年 2 月下旬
营养生长期：约 420 d
初花期：2023 年 3 月下旬
花开放时间：约 21 d
成熟期：2023 年 7 月上旬
果实发育期：约 100 d
果实形状：短圆筒形
单果质量：452.3 g
纵径：7.6 cm
横径：9.7 cm
果形指数：0.8
未成熟果实果皮颜色：银绿色
成熟果实果皮颜色：金黄 / 鲜黄色
果颈：无
果基：平
果顶：平
果实小果能否剥离：不可剥离
果眼外观：微隆起
果眼深度：深
果眼排列方式：左旋
果瘤：无
果肉颜色：淡黄色
果实香味：清香 / 微香
果实风味：酸甜
果肉质地：脆 / 爽脆
果实外观综合评价：中
果实品质综合评价：中
果肉可溶性固形物含量：12.5%
果实成熟特性：晚熟
成熟期的一致性：不一致
丰产性：低产

Moris Gajah

Code: ACCESSION000077
Name: Moris Gajah
Place of origin: Malaysia
Resource type: introduced variety
Main ways of consumption: fresh or processing
Initially introduction place: SSCRI, CATAS introduced from Malaysia in 2010
Plant posture: spreading
Planting time: early January in 2022
Open heart stage: late February in 2023
Vegetative growth period: approx. 420 d
Initial flowering time: late March in 2023
Flowering period: approx. 21 d
Fruit maturation time: early July in 2023
Fruit developing period: approx. 100 d
Fruit shape: short cylindrical
Single fruit weight: 452.3 g
Longitudinal diameter: 7.6 cm
Transverse diameter: 9.7 cm
Fruit shape index: 0.8
Immature fruit peel color: silver green
Mature fruit peel color: golden yellow/vivid yellow
Fruit neck: absent
Fruit base: flat
Fruit top: flat
Fruitlets adhesion situation: inseparable
Fruit eyes appearance: slightly embossed
Fruit eyes depth: deep
Fruit eyes arrangement: levorotation
Fruit tumors: absent
Flesh color: pale yellow
Fruit aroma: faint/slight scent
Fruit flavour: sour-sweet
Flesh texture: crisp/crunchy
Fruit appearance comprehensive evaluation: medium
Fruit quality comprehensive evaluation: medium
Soluble solids content in flesh: 12.5%
Fruit ripening characteristics: late mature
Maturity consistency: nonuniform
Productivity: low

植株
Plant

现红
Open heart

花
Flower

花序
Inflorescence

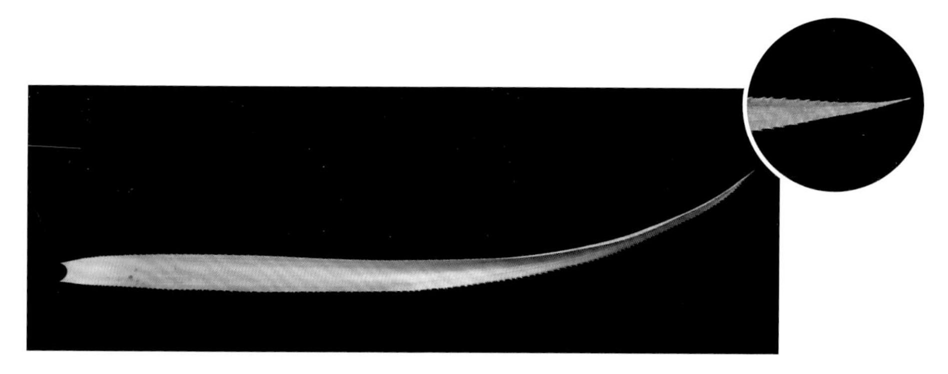

叶片
Leaf

带冠芽果
Fruit with crown bud

冠芽叶刺
Leaf spines in crown bud

果实
Fruit

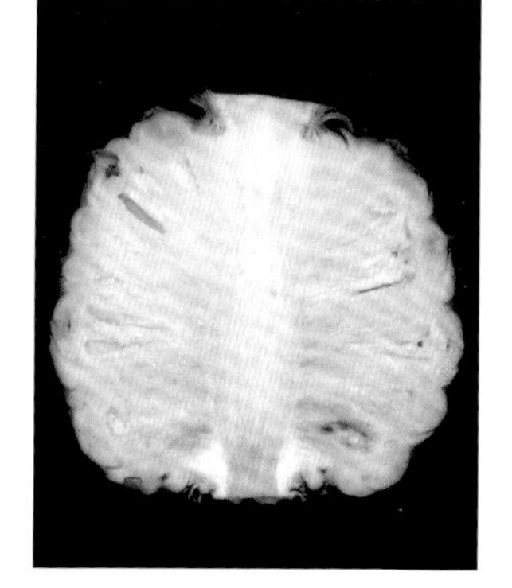

果实纵切
Fruit longitudinal section

果实横切
Fruit transverse section

参考文献 Reference

陈业渊，2005. 热带、南亚热带果树种质资源描述规范 [M]. 北京：中国农业出版社.

李渊林，曾小红，孙光明，2008. 国外菠萝品种资源 [J]. 世界农业，345(1): 55-58.

李渊林，孙光明，2007. 台湾主要商业凤梨品种 [J]. 热带农业科学，27(6): 43-45.

徐迟默，杨连珍，2007. 菠萝科技研究进展 [J]. 华南热带农业大学学报，13(1): 24-29.

中华人民共和国农业部，2015. NY/T 2813—2015 热带作物种质资源描述规范 菠萝 [S]. 北京：中国农业出版社.

D'EECKENBRUGGE G C, LEAL F, DUVAL M F, 2010. Germplasm resources of pineapple[J]. Horticultural Reviews, 21: 133-175.